Preparing for the AP® Physics 1 Examination

Barry Panas

to accompany

COLLEGE PHYSICS for the AP® PHYSICS 1 & 2 COURSES

Third Edition

Gay Stewart | Roger Freedman | Todd Ruskell | Philip Kesten

Bedford, Freeman & Worth
High School Publishers

AP® is a trademark registered by the College Board, which is not affiliated with, and does not endorse, this product.

Copyright © 2023, 2019 by Bedford, Freeman & Worth High School Publishers

All rights reserved. No part of this book may be reproduced, stored in a retrieval system, or transmitted in any form or by any means, electronic, mechanical, photocopying, recording, or otherwise, except as may be permitted by law or expressly permitted by the applicable copyright statuses or in writing by the Publisher.

Manufactured in the United States of America.

1 2 3 4 5 6   27 26 25 24

For information, write:
Bedford, Freeman & Worth, 120 Broadway, 25th Floor, New York NY 10271

ISBN-13: 978-1-319-44704-5
ISBN-10: 1-319-44704-X

Table of Contents

Preface	iv
General Overview	1
Chapter 1 – Introduction to Physics	5
Chapter 2 – Linear Motion	18
Chapter 3 – Motion in Two or Three Dimensions	33
Chapter 4 – Forces and Motion I: Newton's Laws	50
Chapter 5 – Forces and Motion II: Applications	71
Chapter 6 – Circular Motion and Gravitation	84
Chapter 7 – Conservation of Energy	97
Chapter 8 – Application of Conservation Principles	113
Chapter 9 – Momentum, Collisions, and Center of Mass	125
Chapter 10 – Rotational Motion I: A New Kind of Motion	141
Chapter 11 – Torque and Rotation II	155
Chapter 12 – Oscillations Including Simple Harmonic	165
Chapter 13 – The Physics of Fluids	181
Chapter 14 – Preparing for the AP® Physics 1 Exam	201
Practice Exam I	208
Practice Exam II	231
Notes	254

Preface

About the Author

Barry Panas is a high school physics teacher at St. John's-Ravenscourt School in Winnipeg, Canada. He has been teaching AP® Physics since 1996 and has graded AP® Physics exams numerous times as an Exam Reader, and has also helped create AP® Physics exam question scoring guides and trained and managed groups of Exam Readers as a Table Leader. Barry has provided AP® Physics teacher training at numerous workshops and summer institutes across the US, Canada, and overseas in his role as a College Board consultant. His teaching is not limited to the classroom as he has provided physics demonstrations and lessons through his YouTube channel "The Physics Dojo."

Read Me First!

AP® Physics 1 is a rewarding course, but many students find it to be challenging. Fortunately, you have two excellent books to help you through it. The textbook **College Physics for the AP Physics 1 & 2 Courses** contains all of the physics content that you need for success on the AP Physics® 1 exam, and the book that you are now reading—**Strive for a 5: Preparing for the AP® Physics 1 Course**—is here to help you to review and get the most out of the textbook.

Most people read most books by starting at the beginning and then reading through every page until they get to the end of the book. That's a pretty good way to read a novel, but few people can learn physics by reading a physics textbook that way. **Learning physics requires more practice and work than simply reading about it**. Let's start by taking a quick "tour" of the textbook itself.

A Tour of Your Textbook

The biggest part of your textbook is the 26 chapters that it contains, but to be clear: **AP® Physics 1 is based on the material found in only chapters 1 to 13** (the rest of the chapters are for AP® Physics 2). We'll take a closer look at these chapters, but first I'd like you to **bookmark the locations of some important sections at the *back* your textbook**. Locate each of the following sections and mark them so that you will be able to find them quickly in the future.

- The last numbered page of the textbook is Page 1356. The very next page is numbered **M1**, and it is the start of the **Math Tutorial**. For now, flip through these pages until you get to the last page of this section (which is page **M20**).

- The **Appendix (A1 to A5)** contains helpful reference information for working on questions and problems in the future. You do not need to memorize the information found here, but familiarize yourself with what kind of information is available.

- The next section is the **Glossary (G1 to G33)**. The glossary provides helpful definitions to important words and terms—in both English and Spanish.

- The **Answers to Odd Problems** section (**ANS1** to **ANS94**) is exactly what it sounds like.

- The last pages of the textbook contain a thorough **Index (I-1 to I-26)** which you should consult whenever you need to quickly locate something in the textbook.

A Tour of a Chapter

All chapters follow the same structure. I'm a fan of energy, so let's use Chapter 7 for this part of the tour:

- This chapter starts on page 286. Notice the box labelled **YOU WILL LEARN TO** on this page. This box lists everything you should be able to do by the end of this chapter.

- Each chapter is broken down into a number of **chapter sections**. Let's take a closer look at one of these sections. Because it has a little bit of everything, let's go to page 297 for section 7-3:

    o Only a relatively small number of equations are noteworthy enough to be considered "formulas." Page 298 introduces 3 formulas—the three equations that have an **Equation in Words** box, and bubbles that point out the meaning of each letter in the equation.

    o Page 300 does not introduce any new equations, but it does have a **Watch Out!** box. These boxes warn you about ideas that are easily misunderstood and can help you to avoid common mistakes.

    o Page 300 also has an **AP® Exam Tip** box—these boxes contain advice that will help you with the AP® Physics 1 exam.

    o Page 301 applies the physics that has just been introduced in a fully worked out example. Worked out examples are easily spotted by their purple shading.

    o The bottom of page 301 summarizes the key ideas from this section in **The Takeaway for Section 7-3**.

    o Page 302 includes a number of questions: **AP® Building Blocks** (easiest level of difficulty), **AP® Skill Builders** (medium level of difficulty), and **AP® Skills in Action** (highest level of difficulty).

- Near the end of the chapter (starting on page 324), a number of pages wrap up the chapter by listing the **Key Terms** and giving a **Chapter Summary**. On page 324 you will find a "What Did You Learn?" box that repeats the same objectives from the start of the chapter, along with which section(s) these objectives were covered in.

- At the very end of the chapter (starting on page 328) are a number of **Review Problems**. **AP® Group Work** is designed to be especially challenging (worthy of working on collaboratively with other students), and **Practice Problems** are designed to be similar in style to those on the actual AP® exam.

At the Start of Each Chapter

Remember that you are not reading a novel; you are trying to reach a deep understanding of physics. **This takes time**. Don't start the next chapter until you have gained functional understanding of the current one.

Start each chapter with **Strive for a 5**:

- Read the **Before you Read the Chapter: Prepare Yourself**. Be sure to follow through with any advice given there, such as reviewing earlier sections.
- Read the **Chapter Overview** for an introduction to what the chapter is all about.
- Read the **Learning Objectives** for the Chapter. Review these as you proceed through the chapter and check them off as you are able.

For Each Section

Work through the chapter one section at a time. Don't start the next section until you have reached a good understanding of the current one.

Strive for a 5	Textbook
Read the **Section Overview**.	
Complete the **While you Read the Section: Important Terms and Equations** while reading the section in the textbook.	Read the section. Remember to keep an eye open for **Watch Out!** boxes and **AP® Exam Tip** boxes. Go through **Examples** carefully and consider working them out yourself on separate paper. Your own solutions should look similar to the provided solutions. Review **The Takeaway** for the section in the textbook.
Do the questions in the **After you Read the Section: Check Your Understanding**. Don't just select letters for multiple choice questions—take the time to write a brief justification for your answer.	
	Do the questions and problems at the end of the section.

At the End of Each Chapter

Strive for a 5	Textbook
Complete the **After you Read the Chapter: Test Yourself** questions.	
	Do the questions and problems at the end of the chapter.

How to Do Well in AP® Physics

Here is my advice on how to maximize your success in AP® Physics 1:

- You have to **study the ideas presented until you thoroughly understand them**. It's not enough to memorize formulas, examples, or anything else. **The exam places considerable emphasis on understanding**, and the exam is designed to test your depth of understanding. Students who have tried to merely "memorize physics" typically do not do well on the exam.

- **You have to get lots of practice with AP® Physics-style questions and problems**. Being good at AP® Physics means being a ninja in the ways of Physics. You don't become a ninja by just reading books about it. You can't just read "you have to be really, really sneaky" and then say to yourself "right—sneaky—got it! That makes sense! I'm now sneaky and one step closer to becoming a ninja!" You have to **practice** sneaking around! And it's not enough to practice sneaking through the exact same room over and over. When it comes to doing your AP® Ninjutsu exam (which unfortunately, is not a real course—yet), you are going to be put into a situation you've never been in before, and you will be expected to demonstrate your "sneakiness" (and all of your other ninja skills) in all kinds of different situations and environments.

- **Learning physics takes time**. Don't expect to be able to learn a lot of material at the last minute. Hopefully you are reading this at least a few months before you take your AP® Physics 1 exam, in which case you are in good shape. Make yourself a study schedule and stick to it. See page 201 for my advice on how to prepare for the AP® exam.

- **As much as possible, eliminate distractions.** Find a quiet place to study. Shut your phone off and keep your laptop and other devices out of reach if you find that you are distracted by them. The more your brain is paying attention to incoming texts or a conversation with a friend across the table, the less your brain is capable of paying attention to learning physics (or anything else for that matter). Multitasking is overrated. Learn how to be a good monotasker.

Chapter 1
Introduction to Physics

Before You Read the Chapter: Prepare Yourself

Make sure that you have the math skills that you will need:

- Read through the **Math Tutorial sections M-1 to M-8** (found right after Page 1356)—and refer back to it whenever you need to do so.

Chapter Overview

Chapter 1 lays the foundation for all future chapters. Take your time to make sure that you get the most out of this chapter as you will be using these ideas regularly throughout all future chapters. Remember that you can also revisit this chapter to review these ideas as needed.

Learning Objectives

- ☐ Explain the roles that concepts and scientific practices play in physics.
- ☐ Describe three key steps in solving any physics problem.
- ☐ Identify the fundamental units used for measuring physical quantities and convert from one set of units to another.
- ☐ Use significant digits in calculations.
- ☐ Use dimensional analysis to check algebraic results.

1-1 Scientists use special practices to understand and describe the natural world

While **You Read the Section: Important Terms and Equations**

Use the space below to define each term in your own words. You may also add any other notes that will be helpful for future review.

theory

law

model

scientific practice

scientific questioning

After You Read the Section: Check Your Understanding

Choose the best answer to each of the following. Use the space provided to write a short justification for your selection. When you're finished, check that you got the right answers for the right reasons!

1. What does physics use to communicate and summarize complicated ideas?
    a. equations
    b. graphs
    c. diagrams
    d. all of these

    Justify your choice:

2. Which one of the following best describes what physics is?
    a. A set of equations that enable us to perform calculations to get answers to problems related to the physical universe
    b. A set of equations that enable us to answer questions about the physical universe
    c. A set of connected ideas and scientific practices that enable us to make sense of the physical universe
    d. A set of mathematical routines and interlinked concepts that must be memorized in order to answer questions and problems related to the physical universe

    Justify your choice:

3. Which one of the following best describes the role of equations used in physics?
    a. Equations are useful tools, but it is important to understand the ideas that the equations represent.
    b. Equations are the tools used to solve problems in physics; all that matters is that you use the correct equation to get the correct answer.
    c. Equations are used all of the time in physics; you get better at physics by memorizing as many equations as you can.
    d. Equations are used to get answers to problems, but nobody understands what they actually mean or how they work.

    Justify your choice:

1-2 Success in physics requires well-developed problem solving using mathematical, graphical, and reasoning skills

While working on physics problems, students are often tempted to take shortcuts in order to get an answer as quickly as possible. This may save time in the short term, but this approach will not help you when problems get harder. **Take the time to learn these steps now, then use them for every physics problem you do in the future!**

While You Read the Section: Important Terms and Equations

This section doesn't introduce any new physics terms or equations; this section **teaches you how to solve physics problems** using three steps: **Set Up**, **Solve**, and **Reflect**. These three steps are vitally important to solving problems in physics and will be used throughout the rest of the textbook!

After You Read the Section: Check Your Understanding

Choose the best answer to each of the following. Use the space provided to write a short justification for your selection. When you're finished, check that you got the right answers for the right reasons!

1. When solving physics problems your goal should be to
   a. get the correct answer as quickly as possible.
   b. solve as much of the problem as you can in your head and then write as little as possible on your page (preferably just the answer).
   c. locate the correct equation, plug the numbers given in the problem into the equation, and then calculate the answer to the problem.
   d. solve the problem and communicate your solution clearly on your page, and then reflect on the problem to deepen your understanding of physics and sharpen your problem-solving skills.

   Justify your choice:

2. Solving a physics problem
   a. may include numbers, calculations, graphs, diagrams, and written words and sentences as needed.
   b. should include only numbers, calculations, graphs, and diagrams if needed.
   c. should include only numbers, calculations, and graphs if needed.
   d. should include only numbers and calculations.

   Justify your choice:

1-3 **Scientists use simplifying models to make it possible to solve problems; an "object" will be an important model in your studies**

This is a short but important section! In particular, it is important that you understand how the words **"object"** and **"system"** will be used from this point forward. For example, the word "object" is not being used here the way it is usually used outside of physics. It is likely that you won't fully appreciate this until several chapters from now, but don't worry—I'll remind you to review this section often!

While You Read the Section: Important Terms and Equations

Use the space below to define each term in your own words. You may also add any other notes that will be helpful for future review.

fundamental particles

object

object model

system

system model

Chapter 1 | Introduction to Physics

After You Read the Section: Check Your Understanding

Choose the best answer to each of the following. Use the space provided to write a short justification for your selection. When you're finished, check that you got the right answers for the right reasons!

1. Which one of the following is best described as being a fundamental particle?

    a. A speck of dust

    b. An electron

    c. An atom

    d. A molecule

    Justify your choice:

2. In which one of the following situations would the object model be reasonably used for a car?

    a. When calculating how far a car will move when moving at a certain speed for a certain amount of time.

    b. When calculating how many times one of the car's tires will rotate when moving at a certain speed for a certain amount of time.

    c. When calculating how much the springs in the car will compress when weight is added to the car.

    d. When designing a car to maximize the safety of people in the car.

    Justify your choice:

3. Which one of the following best explains the difference between the object model and the system model?

    a. The object model should always be used for items such as a single block, and the system model should always be used for groups of items (such as a group of two blocks).

    b. The system model should always be used for items such as a single block, and the object model should always be used for groups of items (such as a group of two blocks).

    c. The object model should be used when details (such as internal structure, changes in shape, or rotation) can be ignored when solving the problem.

    d. The system model should be used when details (such as internal structure, changes in shape, or rotation) can be ignored when solving the problem.

    Justify your choice:

1-4 Measurements in physics are based on standard units of time, length, and mass

This section introduces **units** and unit conversions. This is very important, so take the time to learn how to **convert units properly** using the method shown in this section.

Table 1-1 on page 10 introduces the **7 fundamental quantities and their SI Units**. We will be working closely with only the first three of these fundamental units in AP® Physics 1, so **it is worth memorizing the units and abbreviations for time, length, and mass**.

While You Read the Section: Important Terms and Equations

Use the space below to define each term in your own words. You may also add any other notes that will be helpful for future review.

unit

Système International (SI)

mass

fundamental units

scientific notation

exponent

After You Read the Section: Check Your Understanding

Choose the best answer to each of the following. Use the space provided to write a short justification for your selection. When you're finished, check that you got the right answers for the right reasons!

1. Which of the following is a set of SI units that can be combined in a number of ways to measure a wide range of physical quantities?
    a. miles, hours, and pounds
    b. meters, seconds, and kilograms
    c. kilometers, seconds, and grams
    d. meters, seconds, and grams

   Justify your choice:

2. Which one of the following shows the correct way to convert a time of two minutes into seconds?
    a. $2 \text{ min} (60) = 120 \text{ s}$
    b. $2 \text{ min} (60 \text{ s}) = 120 \text{ s}$
    c. $2 \text{ min} \left(\dfrac{60 \text{ s}}{1 \text{ min}} \right) = 120 \text{ s}$
    d. $2 \text{ min} \left(\dfrac{1 \text{ min}}{60 \text{ s}} \right) = 120 \text{ s}$

   Justify your choice:

3. Which one of the following is equal to a distance of 150 km?
    a. 1.5×10^2 m
    b. 1.5×10^3 m
    c. 1.5×10^4 m
    d. 1.5×10^5 m

   Justify your choice:

1-5 Correct use of significant digits helps keep track of uncertainties in numerical values

While You Read the Section: Important Terms and Equations

This section introduces only one term, but this term is related to the important concept of **uncertainty**. AP® Physics utilizes significant digits to address uncertainty in most "pen-and-paper problems," but it is worth pointing out that in an experimental setting, uncertainty can be addressed in other ways. This will be further developed in later chapters. Use the space below to define the term in your own words. You may also add any other notes that will be helpful for future review.

significant digits

After You Read the Section: Check Your Understanding

Choose the best answer to each of the following. Use the space provided to write a short justification for your selection. When you're finished, check that you got the right answers for the right reasons!

1. A student measures the length of a block and reports the length to be 12 inches. In this context, the value of "12 inches" has how many significant digits?

   a. none
   b. 1
   c. 2
   d. infinite

   Justify your choice:

2. The unit known as a foot is exactly equal to 12 inches. In this context, the value of "12 inches" has how many significant digits?

   a. none
   b. 1
   c. 2
   d. infinite

   Justify your choice:

3. Two students measure the length of the same object. One student reports the length to be 12 cm while the other student reports the length to be 12.2 cm. Assuming that both students are using significant digits correctly, which of the following best explains the difference in these two measurements?

   a. The measurement reported as 12.2 cm has been taken more carefully and so contains more precision.

   b. The measurement reported as 12.2 cm has been taken more carefully and so contains more uncertainty.

   c. The students disagree on the length of the object since one student is reporting a greater length.

   d. There is no difference since 12.2 cm rounds off to 12 cm.

   Justify your choice:

4. Which of the following gives the result to the following calculation using the proper number of significant digits? $2000.0 + 400 + 80 + 5$

   a. 2000   b. 2500   c. 2490   d. 2485

   Justify your choice:

5. Which of the following gives the result to the following calculation using the proper number of significant digits? 3.000×4.00

   a. 10   b. 12   c. 12.0   d. 12.00

   Justify your choice:

6. Which of the following gives the result to the following calculation using the proper number of significant digits? 50.00×30.0

   a. 1500

   b. 1500.0

   c. 1.5×10^3

   d. 1.50×10^3

   Justify your choice:

1-6 Dimensional analysis is a powerful way to check the results of a physics calculation

Don't be intimidated by the term "Dimensional Analysis" ... it's actually a pretty straight-forward way to inspect equations and can be helpful for identifying errors!

While You Read the Section: Important Terms and Equations

Use the space below to define each term in your own words. You may also add any other notes that will be helpful for future review.

dimension

dimensional analysis

After You Read the Section: Check Your Understanding

Choose the best answer to each of the following. Use the space provided to write a short justification for your selection. When you're finished, check that you got the right answers for the right reasons!

1. Energy has dimensions of $\frac{\text{mass} \times \text{length}^2}{\text{time}^2}$ and speed v has dimensions of $\frac{\text{length}}{\text{time}}$. Which one of the following expressions has the same dimensions as energy?

   a. m/v b. mv c. m^2v d. mv^2

   Justify your choice:

2. Mass m has dimensions of mass, distance r has dimensions of length and force F has dimensions of $\frac{\text{mass} \times \text{length}}{\text{time}^2}$. In the equation $F = \frac{Gm_1m_2}{r^2}$, what are the dimensions of G?

   a. $\frac{\text{length}^3}{\text{mass} \times \text{time}^2}$ b. $\frac{\text{length}^2}{\text{mass} \times \text{time}^2}$ c. $\frac{\text{mass} \times \text{length}^3}{\text{time}^2}$ d. $\frac{\text{mass} \times \text{length}^2}{\text{time}^2}$

   Justify your choice:

After You Read the Chapter: Test Yourself

After reading the chapter and trying the questions below, I recommend that you then **work on the Chapter 1 Review Problems** at the end of the chapter in your textbook before continuing into the next chapter.

End of Chapter Multiple Choice Questions

1. Which one of the following is NOT one of the seven Fundamental Quantities?

    a. weight   b. temperature   c. length   d. time

2. What are the 3 steps used for problem solving, along with an accurate description of each step?

    a. Set Up—get pen, paper, and calculator and prepare your workspace
       Solve—get the answer to the problem
       Reflect—check your answer to verify that you got it correct

    b. Set Up—select the concepts, models, and representations needed in the problem
       Solve—work through the problem doing the math or making and justifying predictions
       Reflect—think about your answer to ensure that it makes sense

    c. Select—choose the equation needed to solve the problem
       Solve—get the answer to the problem
       Reflect—check your answer to verify that you got it correct

    d. Select—choose the equation needed to solve the problem
       Solve—get the answer to the problem
       Reflect—solve the problem a second time and check that your two answers agree with each other

3. Which one of the following is the same speed as 75.0 mi/h? Refer to page A2 of the Appendix at the back of your textbook for conversion information.

    a. 20.8 m/s   b. 33.5 m/s   c. 46.9 m/s   d. 121 m/s

4. Which one of the following is the same area as 2.00 acres? Refer to page A2 of the Appendix at the back of your textbook for conversion information.

    a. 4050 m^2 b. 8090 m^2 c. $26\,600 \text{ m}^2$ d. $3.28 \times 10^7 \text{ m}$

5. Which one of the following is equal to the radius of the Moon? Refer to page A2 of the Appendix at the back of your textbook for conversion information and Page A3 for Astronomical Data.

    a. 2.79×10^6 mi   b. 2.79×10^9 mi   c. 1.08×10^3 mi   d. 1.08×10^6 m

6. What are the dimensions of the Boltzmann constant? Refer the back of your textbook for **Physical Constants** on page A4, **Derived Units** on page A1, and **Fundamental Quantities and their SI Units** in table 1-1 on page 10.

    a. $\dfrac{\text{mass} \times \text{length}}{\text{temp} \times \text{time}}$ b. $\dfrac{\text{mass}^2 \times \text{length}}{\text{temp}^2 \times \text{time}}$ c. $\dfrac{\text{mass} \times \text{length}^2}{\text{temp}^2 \times \text{time}}$ d. $\dfrac{\text{mass} \times \text{length}^2}{\text{temp} \times \text{time}^2}$

End of Chapter Problems

Solve each of the following problems on separate paper.

1. The average "x_{ave}" of a set of 4 numbers can be found by adding the numbers and then dividing by 4. The numbers themselves can be referred to as x_1, x_2, x_3, x_4. If the average of 4 numbers is 49.5, and the first three numbers are 28.0, 42.0 and 71.0, then what is the fourth number?

2. According to the Pythagorean theorem, the lengths of the two legs of a right triangle (referred to as "a" and "b") can each be squared and then added together to get the square of the length of the hypotenuse (referred to as "c"). What is the length of the hypotenuse of a triangle that has legs of length 2.50 m and 3.80 m?

3. If a right triangle has a hypotenuse of length 53.0 m and a leg of length 45.0 m, what is the length of its other leg?

4. The area of a square is 350 m². What is the perimeter of this square?

Chapter 2
Linear Motion

Before You Read the Chapter: Prepare Yourself

Be sure that you have a good understanding of:
- The ideas and skills introduced in Chapter 1, much of which will start to be used here.

Chapter Overview

Chapters 2 and 3 together introduce the topic of kinematics—the study of motion without discussing the causes of motion or deeper concepts related to motion (such as momentum and energy, which will be introduced in later chapters).

Students who are new to Physics often find kinematics to be difficult and confusing at first. Fortunately, with time and effort these difficulties usually don't last long. It takes time to get the hang of doing physics, and most initial difficulty occurs with **problem solving** and sorting out initial confusion with the particular way physics uses some common words (like time, speed, and velocity) in ways that differ from their everyday usage.

Take comfort in knowing that **you have two chapters to sort out these ideas, as Chapter 3 will continue to explore many of the same ideas introduced here in Chapter 2**. You should put in a good effort to make sense of Chapter 2 while you are in it, but don't spend too long on this chapter as your understanding of these ideas will naturally increase as you work through future chapters.

Learning Objectives

☐ Define the linear motion of a system and when and how it can be simplified to an object model.

☐ Define and explain the relationships among distance, displacement, instantaneous velocity, average velocity, constant velocity, and speed.

☐ Define acceleration, explain its relationship to velocity, and interpret graphs of velocity versus time.

☐ Create, use, and interpret narrative (written), mathematical, and graphical representations for linear motion with constant acceleration.

☐ Solve constant-acceleration linear motion problems when given a description of the motion of an object. The description could be in the form of words, graphs, or experimental data.

☐ Explain when objects are in free fall (constant acceleration due to gravity).

2-1 Studying linear motion is the first step in understanding physics

This section introduces some important vocabulary terms; it should not take long to read, but before moving on, make sure you understand what these terms mean!

While You Read the Section: Important Terms and Equations

Use the space below to define each term in your own words. You may also add any other notes that will be helpful for future review.

kinematics

linear motion

After You Read the Section: Check Your Understanding

Choose the best answer to each of the following. Use the space provided to write a short justification for your selection. When you're finished, check that you got the right answers for the right reasons!

1. Which one of the following is an example of linear motion?

   a. An object is thrown upward—it rises for a while, and then falls back down.

   b. An object is thrown to the side—it moves along a curved path to the ground.

   c. A person goes for a walk—they walk to the east for a while, and then walk north for a while.

   d. A person goes for a walk—they walk around a circular path and return to their starting point.

   Justify your choice:

2. If the motion of a car is to be modelled as an object, which part of the car would be best to consider the motion of?

   a. the center of one of its tires

   b. the point at which one of its tires is in contact with the ground

   c. the frontmost part of the car

   d. the middle of the car (more specifically, its center of mass)

   Justify your choice:

Chapter 2 | Linear Motion   19

2-2 Constant velocity means moving at a constant speed without changing direction

This section introduces a lot of important physics vocabulary that can feel overwhelming—especially if you are coming across these ideas for the first time! Take your time with this section and use the **"While you Read"** below to keep track of the meaning of each term as you come across them in the textbook.

This section also introduces two important equations for motion. There will be more equations introduced in later sections, so **use this section to practice solving problems** (this will become more challenging when there are more equations to choose from). Remember that your goal should not be to get the answers as quickly as possible, but rather to reach a deep understanding of the physics in each problem while clearly communicating all steps needed to solve it (Set Up, Solve, Reflect).

While You Read the Section: Important Terms and Equations

Use the space below to define each term in your own words. For equations, use the space to identify what each letter represents and its associated SI unit. You may also add any other notes that will be helpful for future review.

constant velocity

speed

velocity

displacement

coordinate system

origin

20   Section 2-2

position

vector

magnitude

direction

distance

scalar

average velocity

$$v_{\text{average},x} = \frac{x_2 - x_1}{t_2 - t_1} = \frac{\Delta x}{\Delta t}$$

$$x = x_0 + v_x t$$

After You Read the Section: Check Your Understanding

Choose the best answer to each of the following. Use the space provided to write a short justification for your selection. When you're finished, check that you got the right answers for the right reasons!

1. An object is initially at $x = +5.00$ m. It then moves to $x = +10.00$ m and then to $x = +7.00$ m. What is the displacement of the object?

    a. 2.00 m
    b. 3.00 m
    c. 7.00 m
    d. 8.00 m

    Justify your choice:

2. At $t = 2.00$ s an object is located at $x = 10.0$ m. At $t = 5.00$ s the object is at $x = 22.0$ m. What was the object's average velocity for this interval of time?

    a. 3.00 m/s
    b. 4.00 m/s
    c. 4.40 m/s
    d. 6.40 m/s

    Justify your choice:

3. Which of the following position and time data is consistent with an object moving with constant nonzero velocity?

    a. at $t = 0$, $x = 5.00$ m; at $t = 2.00$ s, $x = 5.00$ m; at $t = 3.00$ s, $x = 5.00$ m
    b. at $t = 0$, $x = 1.00$ m; at $t = 2.00$ s, $x = 2.00$ m; at $t = 3.00$ s, $x = 3.00$ m
    c. at $t = 0$, $x = 1.00$ m; at $t = 2.00$ s, $x = 2.00$ m; at $t = 4.00$ s, $x = 2.50$ m
    d. at $t = 0$, $x = 1.00$ m; at $t = 2.00$ s, $x = 2.00$ m; at $t = 4.00$ s, $x = 3.00$ m

    Justify your choice:

4. The slope of an x-t graph is equal to which one of the following?

    a. initial position
    b. end position
    c. average position
    d. velocity

    Justify your choice:

5. At $t = 0$ an object is at $x = 15.0$ m. If the object moves with a constant velocity of 5.00 m/s, where will it be at $t = 5.00$ s?

    a. 15.0 m
    b. 25.0 m
    c. 40.0 m
    d. 65.0 m

    Justify your choice:

2-3 Velocity is the rate of change of position, and acceleration is the rate of change of velocity

This section introduces **acceleration**, including the equation that defines it. Make every effort to achieve a good understanding of acceleration before moving onto the next section, as acceleration will be central to the rest of the chapter, and will continue to be an important concept in future chapters.

While You Read the Section: Important Terms and Equations

Use the space below to define each term in your own words. For equations, use the space to identify what each letter represents and its associated SI unit. You may also add any other notes that will be helpful for future review.

acceleration

instantaneous

$$a_{\text{average},x} = \frac{\Delta v_x}{\Delta t} = \frac{v_{2x} - v_{1x}}{t_2 - t_1}$$

Chapter 2 | Linear Motion 23

After You Read the Section: Check Your Understanding

Choose the best answer to each of the following. Use the space provided to write a short justification for your selection. When you're finished, check that you got the right answers for the right reasons!

1. Given the x-t graph for an accelerating object, which of the following would give the instantaneous velocity at $t = 5.0$ s?

    a. The displacement of the object at $t = 5.0$ s divided by 5.0 s

    b. The velocity at $t = 5.0$ s divided by 5.0 s

    c. The slope of the graph at $t = 5.0$ s

    d. The average slope of the graph between $t = 0$ and $t = 5.0$ s

    Justify your choice:

2. At $t_1 = 1.00$ s an object's velocity is 10.0 m/s. At $t_2 = 5.00$ s the object's velocity is 20.0 m/s. What was the object's average acceleration for this time interval?

    a. 2.50 m/s^2 b. 4.00 m/s^2 c. 5.00 m/s^2 d. 7.50 m/s^2

    Justify your choice:

3. Which of the following is equal to the instantaneous acceleration a_x of an object at $t = t_1$?

    a. the slope of its x-t graph at $t = t_1$

    b. the slope of its v_x-t graph at $t = t_1$

    c. its position x_1 at $t = t_1$ divided by time t_1

    d. its velocity v_{x1} at $t = t_1$ divided by time t_1

    Justify your choice:

4. An object has negative velocity v_x and negative acceleration a_x. Which of the following best describes its motion?

    a. It is moving in the $+x$ direction while slowing down

    b. It is moving in the $+x$ direction while speeding up

    c. It is moving in the $-x$ direction while slowing down

    d. It is moving in the $-x$ direction while speeding up

    Justify your choice:

2-4 Tools for describing constant acceleration motion

This section continues to explore the relationships among the kinematic quantities (time, position, velocity, and acceleration). It also introduces three more kinematics equations which collectively may be referred to as "**THE Kinematics Equations**" for acceleration motion. These three equations will get used a lot throughout the rest of the chapter and beyond, so take the time to get comfortable with them now.

While You Read the Section: Important Terms and Equations

Use the space below to define each term in your own words. For equations, use the space to identify what each letter represents and its associated SI unit. You may also add any other notes that will be helpful for future review.

$$v_x = v_{0x} + a_x t$$

$$x = x_0 + v_{0x}t + \frac{1}{2}a_x t^2$$

$$v_x^2 = v_{0x}^2 + 2a_x(x - x_0)$$

After You Read the Section: Check Your Understanding

Choose the best answer to each of the following. Use the space provided to write a short justification for your selection. When you're finished, check that you got the right answers for the right reasons!

1. Which of the following velocity and time data is consistent with an object moving with constant nonzero acceleration?

    a. at $t = 0$, $v_x = 5.00$ m/s; at $t = 1.00$ s, $v_x = 10.00$ m/s; at $t = 2.00$ s, $v_x = 20.00$ m/s
    b. at $t = 0$, $v_x = 5.00$ m/s; at $t = 1.00$ s, $v_x = 10.00$ m/s; at $t = 2.00$ s, $v_x = 15.00$ m/s
    c. at $t = 0$, $v_x = 5.00$ m/s; at $t = 1.00$ s, $v_x = 10.00$ m/s; at $t = 2.00$ s, $v_x = 25.00$ m/s
    d. at $t = 0$, $v_x = 10.00$ m/s; at $t = 4.00$ s, $v_x = 20.00$ m/s; at $t = 6.00$ s, $v_x = 30.00$ m/s

    Justify your choice:

2. Which of the following best describes the ***x-t* graph** of an object that is initially at rest and then moves with positive acceleration?

    a. a parabola that curves upward
    b. a parabola that curves downward
    c. a straight line with positive slope
    d. a straight line with negative slope

    Justify your choice:

3. Which of the following best describes the ***v_x-t* graph** of an object that is initially at rest and then moves with positive acceleration?

    a. a parabola that curves upward
    b. a parabola that curves downward
    c. a straight line with positive slope
    d. a straight line with negative slope

    Justify your choice:

4. Which of the following is equal to displacement of an object?

    a. slope of an *x-t* graph
    b. slope of a *v_x-t* graph
    c. area under an *x-t* graph
    d. area under a *v_x-t* graph

    Justify your choice:

Section 2-4

2-5 Solving linear motion problems: constant acceleration

This section extends the ideas introduced in earlier sections by applying them to a wider range of situations, so there are no new terms or equations introduced here. Carefully go over the examples in this section; when you solve problems on your own (including the problems below), you should **aim to have your solutions look similar to the solutions given in the examples.**

After You Read the Section: Check Your Understanding

Solve each of the following problems on separate paper.

1. An object is initially stationary ($v_{0x} = 0$). It then accelerates at $a_x = 5.00$ m/s^2 for $t = 3.00$ s. How fast is it then moving (calculate v_x)?

2. After accelerating at $a_x = 2.00$ m/s^2 for $t = 12.0$ s, an object is moving at $v_x = 76.0$ m/s. How was it initially moving (calculate v_{0x})?

3. An object is initially stationary ($v_{0x} = 0$) at the origin ($x = 0$). It then accelerates at $a_x = 1.50$ m/s^2. How long will it take (calculate t) for the object to reach $x = 108$ m?

4. A car was initially moving on the highway at $v_{0x} = 8.00$ m/s. It then accelerates at $a_x = 4.00$ m/s^2. What is the displacement of the car in the time it takes it to reach a speed of 20.0 m/s?

2-6 Objects falling freely near Earth's surface have constant acceleration

Although 3 "new" equations are introduced in this section, it is helpful to realize that these are nearly the same three equations used in the previous two sections. The only difference is that the equations in this section are specifically for the vertical motion of an object that is being affected by only gravity; the x's in the previous set of equations are replaced with y's here, and a_x is replaced with $-g$ (upward is taken as positive).

While You Read the Section: Important Terms and Equations

Use the space below to define each term in your own words. For equations, use the space to identify what each letter represents and its associated SI unit. You may also add any other notes that will be helpful for future review.

free fall

acceleration due to gravity

$$v_y = v_{0y} - gt$$

$$y = y_0 + v_{0y}t - \frac{1}{2}gt^2$$

$$v_y^2 = v_{0y}^2 - 2g(y - y_0)$$

After You Read the Section: Check Your Understanding

Choose the best answer to each of the following. Use the space provided to write a short justification for your selection. When you're finished, check that you got the right answers for the right reasons!

1. Which one of the following is the best example of an object in free fall?

    a. A rock is dropped at the water's surface in a swimming pool; it falls to the bottom of the pool.

    b. A feather is dropped; it slowly falls to the ground.

    c. A rock is thrown upward; it rises from the point of release to the highest point it reaches.

    d. A block of wood is released from the bottom of a pool; it rises to the surface.

    Justify your choice:

2. Which one of the following best describes the magnitude of the free-fall acceleration, g?

    a. It is positive when an object moves upward, and negative when an object moves downward.

    b. It is negative when an object moves upward, and positive when an object moves downward.

    c. It is always negative.

    d. It is always positive.

    Justify your choice:

3. Which one of the following best describe when a value of $g = 9.8$ m/s^2 can be used?

    a. Any time that an object moves vertically.

    b. Only when an object is in free-fall near Earth's surface.

    c. Any time that an object is in free-fall anywhere in the universe.

    d. Only when an object moves with negligible air resistance.

    Justify your choice:

After You Read the Chapter: Test Yourself

After reading the chapter and trying the questions below, I recommend that you then **work on the Chapter 2 Review Problems** at the end of the chapter in your textbook before continuing into the next chapter.

Word Match

A) Kinematics B) Linear Motion C) Coordinate System D) Origin
E) Vector F) Scalar G) Time H) Time Interval
I) Position J) Displacement K) Distance L) Average Velocity
M) Instantaneous N) Average Speed O) Motion Diagram P) Rate of Change
Q) Acceleration R) Magnitude S) Position-Time Graph T) Velocity-Time Graph
U) Free Fall

___ 1. Determined by comparing two clock readings, this is also known as elapsed time.

___ 2. A measure of how quickly the velocity of an object is changing.

___ 3. Motion in which the only two available directions to move in are opposite of each other.

___ 4. This necessarily changes when an object has a non-zero displacement.

___ 5. For an object that moves around a circle, this would be equal to the circle's circumference.

___ 6. The slope of this would be zero for an object moving with constant velocity.

___ 7. Displacement and velocity are examples of this; distance and speed are not.

___ 8. The study of motion including velocity and acceleration.

___ 9. Can be found by the slope of a secant line on a position-time graph.

___ 10. Another word for size.

___ 11. Used to indicate a value at a specific moment in time.

___ 12. Although much less precise, this could be a "calendar reading."

___ 13. On a position-time graph, this would be the "rise."

___ 14. If a car dripped one drop of oil every second, it would produce a very large one of these.

___ 15. An indication of how quickly two variables change in relation to each other.

___ 16. Temperature and the cost of an item are examples of this kind of measurement.

___ 17. For a walk around the block, find this by dividing the block's perimeter by elapsed time.

___ 18. Relocating this would change the position of everything, but not their displacements.

___ 19. If this is curved, then the object was accelerating; if it is straight, it wasn't accelerating.

___ 20. Ideally this requires there to be no air, but we'll accept "negligible" influence from air.

___ 21. Defining this is necessary in order to use signs to indicate directions.

End of Chapter Multiple Choice Questions

1. An object moves in a line along the x-axis. Which of the following correctly compares the magnitudes of its ending position, x, its displacement, and the distance it moved?
   a. distance $\geq$ displacement; x could be anything
   b. distance $\geq$ displacement $\geq x$
   c. distance $\geq$ displacement $= x$
   d. distance $=$ displacement $= x$

2. Beginning at time $t = 0$, an object moves a distance d_1 with constant speed v_1. At time $t = t_1$ its speed quickly changes to a different speed v_2 in a negligible amount of time. It continues to move with constant speed v_2 for an additional distance d_2, which it completes at time $t = t_2$. For the motion of the object over the interval from $t = 0$ to $t = t_2$, which of the following is a correct expression for the object's average speed?
   a. $\dfrac{d_1 + d_2}{t_1 + t_2}$
   b. $\dfrac{v_1 t_1 + v_2 t_2}{t_1 + t_2}$
   c. $\dfrac{d_1 + d_2}{t_2}$
   d. $\dfrac{v_1 + v_2}{2}$

3. Which of the following is necessarily true for an object that moves in a line with constant acceleration?
   a. Its velocity must change by the same amount in any two equal intervals of time.
   b. Its position must change by the same amount in any two equal intervals of time.
   c. Its displacement must change by the same amount in any two equal intervals of time.
   d. Its distance must change by the same amount in any two equal intervals of time.

For the next two questions: The graph shown here may be a position-time graph or a velocity-time graph for an object that moves in a line.

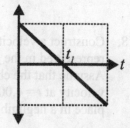

4. If the graph is a **position-time graph**, which of the following best describes the motion of the object it represents?
   a. The object slowed down until time $t = t_1$, and then sped up.
   b. The object moved with constant speed for the entire interval of time shown.
   c. The object reversed the direction of its motion at time $t = t_1$.
   d. The object reversed the direction of its motion at $t = 0$.

5. If the graph is a **velocity-time graph**, which of the following best describes the motion of the object it represents?
   a. The object slowed down for the entire interval of time shown.
   b. The object moved with constant velocity for the entire interval of time shown.
   c. The object reversed the direction of its motion at time $t = t_1$.
   d. The object moved in the same direction for the entire interval of time shown.

End of Chapter Problems

Solve each of the following problems on separate paper.

1. A girl begins at the western-most point of a circular track that has a radius of 100.0 m. She then walks halfway around the track. Find her displacement.

2. An object moves 20.0 meters in 10.0 seconds. What was its average speed?

3. A car moves at a constant speed of 10.0 m/s for 2.00 minutes. It then suddenly changes to a new speed of 20.0 m/s which it then maintains for an additional 3.00 minutes. For the entire motion described, find the car's average speed.

4. A car begins from rest. It moves with a constant acceleration of 2.00 m/s². Find how far it moves in 5.00 seconds.

5. A long smooth track is tilted: one end is higher than the other. A cart is on the track and sent toward the top end of the track with an initial speed of 8.00 m/s. The cart moves with constant acceleration on the track, initially slowing down at a rate of 4.00 m/s². What is the average velocity of the cart's first 5.00 seconds of motion, beginning from the moment it is released?

6. A ball is thrown upwards at 15.0 m/s. How long does it take the ball to reach its highest point?

7. A ball is dropped. Find how far it falls in 3.00 seconds.

8. Construct a velocity-time graph for the motion represented in the given position-time graph. Assume that the changes to the object's velocity at $t = 4.00$ s and at $t = 8.00$ s take place in a negligible amount of time.

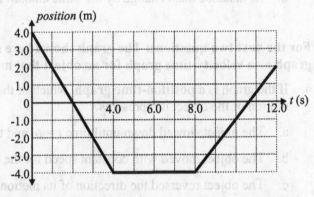

9. Construct a position-time graph for the motion represented in the given velocity-time graph, given that the object was at a position of -10 m at $t = 0$.

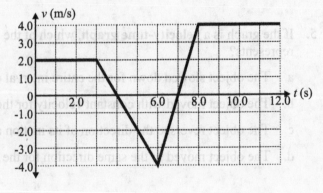

Chapter 3
Motion in Two or Three Dimensions

Before You Read the Chapter: Prepare Yourself

Be sure that you have a good understanding of:

- **Chapter 2**—all of it!
- **Trigonometry**—review M-8 from the Math Tutorial at the back of your textbook.

Chapter Overview

Chapter 3 applies the ideas and equations from Chapter 2 to the motion of an object in two (or even three) dimensions. The motion of a **projectile** (an object that moves while being influenced only by gravity) in two dimensions will be especially important in this chapter. The chapter begins, however, with a more in-depth exploration of **vectors**, which are essential in this and future chapters.

Learning Objectives

- ☐ Recognize the differences among one-, two-, and three-dimensional motion.
- ☐ Describe the properties of a vector and how to find the sum or difference of two vectors.
- ☐ Describe the motion of an object in two dimensions using quantities such as displacement, distance, velocity, speed, and acceleration, in an appropriate way for a chosen coordinate system.
- ☐ Explain how the displacement, velocity, and acceleration of an object or the center of mass of a system are described in terms of vectors, and can be modeled mathematically by the kinematic equations when acceleration is constant.
- ☐ Recognize that the velocity observed for an object depends on the reference frame in which it is observed, and be able to describe the reference frames of a given observer from a description of the physical situation.
- ☐ Be able to convert the velocity of an object relative to one reference frame into the velocity of that same object observed from a different reference frame.
- ☐ Identify the key features of projectile motion and how to interpret this kind of motion.
- ☐ Solve problems involving projectile motion.

3-1 The ideas of linear motion help us understand motion in two or three dimensions

This section introduces a fundamental principle—the **kinematics of linear motion** (motion along one dimension) **can be repeated for each dimension of motion**. This section is introductory and does not introduce any new terms or equations.

After You Read the Section: Check Your Understanding

Choose the best answer to each of the following. Use the space provided to write a short justification for your selection. When you're finished, check that you got the right answers for the right reasons!

1. Which of the following is an example of an object moving in only one dimension?
   a. A block slides down a slanted surface.
   b. An insect runs up a wall with increasing speed.
   c. A feather is dropped from rest—it falls straight down with non-negligible air resistance.
   d. All of the above.

   Justify your choice:

2. When solving a problem involving an object moving in two dimensions, which of the following approaches must be used?
   a. The magnitudes of position, velocity, and acceleration can be used in the familiar kinematics equations, regardless of their directions.
   b. Position, velocity, and acceleration need to be treated as separate one-dimensional problems along each dimension of motion.
   c. Both of the above are true.
   d. None of the above are true.

   Justify your choice:

3-2 A vector quantity has both a magnitude and a direction

Sections 3-2 and 3-3 are closely connected. Both sections are unusual in that they are not just about "physics" ... instead, they introduce how to work with **vectors** in a way that will be used extensively in the rest of the chapter and beyond. **It is important to gain a good understanding of vectors for success in the rest of the course, so take your time with sections 3-2 and 3-3!** Most of the details will come in the next section; Section 3-2 gives only an overview of what vectors are all about, and introduces how they can be added, subtracted, and multiplied by scalars.

While You Read the Section: Important Terms and Equations

Use the space below to define each term in your own words. You may also add any other notes that will be helpful for future review.

vector addition

vector difference

After You Read the Section: Check Your Understanding

Choose the best answer to each of the following. Use the space provided to write a short justification for your selection. When you're finished, check that you got the right answers for the right reasons!

1. Which of the following is the main difference between vectors and scalars?

    a. Scalars can be positive or negative; vectors can never be negative.

    b. Vectors can be positive or negative; scalars can never be negative.

    c. Scalars have a magnitude and a direction; vectors do not have a direction.

    d. Vectors have a magnitude and a direction; scalars do not have a direction.

    Justify your choice:

2. Which of the following correctly describes a way that two vectors can be added?

    a. Place the tail of the second vector at the tip of the first vector; the sum points from the tail of the first vector to the tip of the second vector.

    b. Place the tail of the second vector at the tip of the first vector; the sum points from the tip of the first vector to the tail of the second vector.

    c. Place the tail of the second vector at the tail of the first vector; the sum points from the tip of the first vector to the tip of the second vector.

    d. Place the tail of the second vector at the tail of the first vector; the sum points from the tip of the second vector to the tip of the first vector.

    Justify your choice:

3. Which of the following correctly describes a way that vector $\vec{F}$ can be subtracted from vector $\vec{G}$ (i.e., $\vec{G} - \vec{F}$)?

    a. Make a new vector that has the same magnitude as $\vec{G}$, but points in the opposite direction. The difference is found by adding this new vector to $\vec{F}$.

    b. Make a new vector that has the same magnitude as $\vec{F}$, but points in the opposite direction. The difference is found by adding this new vector to $\vec{G}$.

    c. Make two new vectors by reversing the directions of both $\vec{F}$ and $\vec{G}$. The difference is found by adding the two new vectors.

    d. Add $\vec{F}$ and $\vec{G}$ to get their sum. The difference is the new vector that has the same magnitude as this sum, but with the opposite direction.

    Justify your choice:

4. Vector $\vec{H}$ has a direction of 30° north of east. Which one of the following is the direction of $-\vec{H}$?

   a. 30° south of west
   b. 30° west of south
   c. 30° east of north
   d. -30° east of north

   Justify your choice:

5. Vector $\vec{I}$ has a direction of 25° west of north. Which of the following is another way to indicate this same direction?

   a. 25° north of west
   b. 25° east of south
   c. 65° west of north
   d. 65° north of west

   Justify your choice:

6. Vector $\vec{J}$ has magnitude 3 and direction 10° north of east. Scalar $k = -2$. Which of the following is equal to the product $k\vec{J}$?

   a. -6, -20° north of east
   b. -6, 20° north of east
   c. 6, 10° south of west
   d. 6, -10° north of east

   Justify your choice:

3-3 Vectors can be described in terms of components

Section 3-3 is a direct continuation of Section 3-2. Adding vectors, subtracting vectors, and multiplying a vector by a scalar are covered again, but this time using a **vector component** approach. Future sections and chapters will make regular use of vector components, so make every effort to reach a good understanding of them here.

While You Read the Section: Important Terms and Equations

Use the space below to define each term in your own words. For equations, use the space to identify what each letter represents and its associated SI unit. Carefully define from which axis angles used in the equations are measured. You may also add any other notes that will be helpful for future review.

component

component method

$$A_x = A\cos\theta$$

$$A_y = A\sin\theta$$

$$A = \sqrt{A_x^2 + A_y^2}$$

$$\tan\theta = \frac{A_y}{A_x}$$

$C_x = A_x + B_x$

$C_y = A_y + B_y$

$D_x = A_x - B_x$

$D_y = A_y - B_y$

$E_x = cA_x$

$E_y = cA_y$

After You Read the Section: Check Your Understanding

Choose the best answer to each of the following. Use the space provided to write a short justification for your selection. When you're finished, check that you got the right answers for the right reasons!

1. Vector $\vec{L}$ has magnitude 15 and is directed at 53.1° west of north. If $+x$ is directed to the east and $+y$ is directed to the north, then which of the following are equal to the components of $\vec{L}$?

   a. $L_x = 9$ west and $L_y = 12$ north
   b. $L_x = 12$ west and $L_y = 9$ north
   c. $L_x = 19$ west and $L_y = 25$ north
   d. $L_x = 25$ west and $L_y = 19$ north

   Justify your choice:

2. Vector $\vec{M}$ has components $M_x = 28$ and $M_y = -45$. If $+x$ is directed to the east and $+y$ is directed to the north, then which of the following is the magnitude of $\vec{M}$?

   a. -17
   b. 17
   c. 53
   d. 73

   Justify your choice:

3. Vector $\vec{M}$ has components $M_x = 28$ and $M_y = -45$. If $+x$ is directed to the east and $+y$ is directed to the north, then which of the following is the direction of $\vec{M}$?

   a. 58° north of east
   b. 32° north of east
   c. 58° south of east
   d. 32° south of east

   Justify your choice:

4. Vector $\vec{N}$ has components $N_x = 8$ m and $N_y = 12$ m. Vector $\vec{P}$ has components $P_x = 3$ and $P_y = -4$. Which of the following are the components of $\vec{Q} = \vec{N} + \vec{P}$?

   a. $Q_x = 11, Q_y = 8$ b. $Q_x = 11, Q_y = 16$ c. $Q_x = 5, Q_y = 8$ d. $Q_x = 5, Q_y = 16$

   Justify your choice:

5. Vector $\vec{N}$ has components $N_x = 8$ m and $N_y = 12$ m. Vector $\vec{P}$ has components $P_x = 3$ and $P_y = -4$. Which of the following are the components of $\vec{Q} = \vec{N} - \vec{P}$?

   a. $Q_x = 11, Q_y = 8$ b. $Q_x = 11, Q_y = 16$ c. $Q_x = 5, Q_y = 8$ d. $Q_x = 5, Q_y = 16$

   Justify your choice:

6. Using a coordinate system in which $+x$ points at $0°$ and $+y$ points at $90°$, a car has velocity $\vec{v}_{car}$ with components $v_{car,x} = 8.0$ m/s and $v_{car,y} = 15$ m/s. A truck has velocity $\vec{v}_{truck} = -2\vec{v}_{car}$. Which of the following is nearest to the velocity of the truck?

   a. 17 m/s directed at $62°$ b. 17 m/s directed at $240°$

   c. 34 m/s directed at $62°$ d. 34 m/s directed at $240°$

   Justify your choice:

3-4 Motion in a plane: reference frames, velocity, and relative motion

This section begins to **apply what was just covered about vectors earlier in this chapter to displacement and velocity from chapter 2**. It also introduces the important idea of **relative motion**, which reveals that things can look very different depending on how you choose to look at and measure them.

While You Read the Section: Important Terms and Equations

Use the space below to define each term in your own words. For equations, use the space to identify what each letter represents and its associated SI unit. You may also add any other notes that will be helpful for future review.

reference frame

trajectory

motion in a plane

relative velocity

$\Delta \vec{r} = \vec{r}_2 - \vec{r}_1$

$\vec{v} = \dfrac{\Delta \vec{r}}{\Delta t} = \dfrac{\vec{r}_2 - \vec{r}_1}{t_2 - t_1}$

$v_x = \dfrac{\Delta x}{\Delta t} = \dfrac{x_2 - x_1}{t_2 - t_1}$

$v_y = \dfrac{\Delta y}{\Delta t} = \dfrac{y_2 - y_1}{t_2 - t_1}$

$\vec{v}_{\text{first, last}} = \vec{v}_{\text{first, other}} + \vec{v}_{\text{other, last}}$

After You Read the Section: Check Your Understanding

Choose the best answer to each of the following. Use the space provided to write a short justification for your selection. When you're finished, check that you got the right answers for the right reasons!

1. An object moves only along the x-axis. It is initially at position $x = 5$ m. It then moves to $x = 20$ m, and then to $x = -10$ m. For the motion described, what is the object's displacement?

    a. -10 m   b. -15 m   c. 15 m   d. 45 m

    Justify your choice:

2. An object moves around a circular path. Which of the following correctly describes the direction of the object's velocity?

    a. it is always directed towards the center of the circle

    b. it is always directed away from the center of the circle

    c. it is always directed along the object's trajectory, tangential to the circle

    d. it is always directed along the object's trajectory, perpendicular to the circle

    Justify your choice:

3. Because of a wind, the air moves relative to the ground with velocity $\vec{v}_{AG}$. A plane flies with velocity $\vec{v}_{PA}$ relative to the air. Which of the following is equal to the velocity of the plane relative to the ground?

    a. $\vec{v}_{AG} + \vec{v}_{PA}$ b. $\vec{v}_{PA} + \vec{v}_{AG}$ c. $\vec{v}_{AG} - \vec{v}_{PA}$ d. $\vec{v}_{PA} - \vec{v}_{AG}$

    Justify your choice:

4. A person swims with the same speed relative to the water, regardless of the direction they swim. The person intends to swim across a river that has a current that flows to the east. If the person is initially on the south side of the river, in which of the following directions should the person swim relative to the water in order cross the river in the shortest time possible?

    a. directly across the river, to the north

    b. in the same direction as the current, to the east

    c. at an angle between north and east

    d. at an angle between north and west

    Justify your choice:

3-5 Motion in a plane: acceleration and projectile motion

This section continues to **apply what was covered about vectors earlier in this chapter to acceleration from Chapter 2** by introducing **projectile motion**. It may appear that several new equations are introduced for projectile motion, but a closer look reveals that these equations have already been introduced.

While You Read the Section: Important Terms and Equations

Use the space below to define each term in your own words. For equations, use the space to identify what each letter represents and its associated SI unit. You may also add any other notes that will be helpful for future review.

projectile motion

projectile

parabola

$$\vec{a} = \frac{\Delta \vec{v}}{\Delta t} = \frac{\vec{v}_2 - \vec{v}_1}{t_2 - t_1}$$

$$a_x = \frac{\Delta v_x}{\Delta t} = \frac{v_{x2} - v_{x1}}{t_2 - t_1}$$

$$a_y = \frac{\Delta v_y}{\Delta t} = \frac{v_{y2} - v_{y1}}{t_2 - t_1}$$

$a_x = 0$

$a_y = -g$

$v_x = v_{0x}$

$v_y = v_{0y} - gt$

$x = x_0 + v_{0x}t$

$y = y_0 + v_{0y}t - \frac{1}{2}gt^2$

After You Read the Section: Check Your Understanding

Choose the best answer to each of the following. Use the space provided to write a short justification for your selection. When you're finished, check that you got the right answers for the right reasons!

1. Which one of the following is NOT a property of an idealized projectile?

    a. it is launched with a nonzero initial velocity

    b. it moves with no air resistance

    c. it is accelerated only by gravity

    d. it is self-propelled (for example, by a motor)

    Justify your choice:

2. At time $t = 0$ a projectile is launched. At $t = 2$ s the object reaches its highest point and at $t = 4$ s it hits the ground. Which of the following best describe the interval(s) of time during which the object was accelerating downward?

    a. only from $t = 0$ s to $t = 2$ s

    b. only from $t = 2$ s to $t = 4$ s

    c. from $t = 0$ s to $t = 4$ s

    d. from $t = 0$ s to just before $t = 2$ s, and again just after $t = 2$ s to $t = 4$ s

    Justify your choice:

3. Which one of the following most accurately describes the components of motion for an object in projectile motion?

    a. constant velocity along the horizontal and changing velocity along the vertical

    b. changing velocity along the horizontal and constant velocity along the vertical

    c. constant velocity along the horizontal and constant velocity along the vertical

    d. changing velocity along the horizontal and changing velocity along the vertical

    Justify your choice:

4. Objects A, B, and C are all initially at the same height above the ground. Object A is dropped and moves for time Δt_A before arriving at the ground. Object B is launched horizontally with a small initial speed and moves for time Δt_B before arriving at the ground. Object C is launched horizontally with a large initial speed and moves for time Δt_C before arriving at the ground. Which of the following correctly compares the amount of time each object takes to reach the ground?

    a. $\Delta t_A > \Delta t_B > \Delta t_C$ b. $\Delta t_A < \Delta t_B < \Delta t_C$ c. $\Delta t_A > \Delta t_B < \Delta t_C$ d. $\Delta t_A = \Delta t_B = \Delta t_C$

    Justify your choice:

3-6 You can solve projectile motion problems using techniques learned for linear motion

This section extends the ideas introduced in earlier sections by applying them to a wider range of situations, so there are no new terms or equations introduced here. Carefully go over the examples in this section; when you solve problems on your own (including the problems below), you should **aim to have your solutions look similar to the solutions given in the examples.**

After You Read the Section: Check Your Understanding

Choose the best answer to each of the following. Use the space provided to write a short justification for your selection. When you're finished, check that you got the right answers for the right reasons!

1. An object is thrown horizontally with an initial velocity of 7.5 m/s from the top of a tall building. Which of the following is closest to its vertical component of velocity one second later?

    a. 7.5 m/s b. 10 m/s c. 12.5 m/s d. 17.5 m/s

    Justify your choice:

2. An object is thrown horizontally with an initial velocity of 7.5 m/s from the top of a tall building. Which of the following is closest to its horizontal component of velocity one second later?

    a. 7.5 m/s b. 10 m/s c. 12.5 m/s d. 17.5 m/s

    Justify your choice:

3. An object is thrown horizontally with an initial velocity of 7.5 m/s from the top of a tall building. Which of the following is closest to the magnitude of its velocity one second later?

    a. 7.5 m/s b. 10 m/s c. 12.5 m/s d. 17.5 m/s

    Justify your choice:

4. An object is thrown horizontally with an initial velocity of 7.5 m/s from the top of a tall building. Which of the following is closest to the angle its velocity vector makes to the horizontal one second later?

    a. 18° b. 27° c. 36° d. 53°

    Justify your choice:

Chapter 3 | Motion in Two or Three Dimensions   47

After You Read the Chapter: Test Yourself

After reading the chapter and trying the questions below, I recommend that you then **work on the Chapter 3 Review Problems** at the end of the chapter in your textbook before continuing into the next chapter.

End of Chapter Multiple Choice Questions

1. Two vectors have magnitudes A and B with $A > B$. Their directions are not known. If the two vectors are added, which of the following is true of the magnitude of their resultant?

    a. It must be between A and B

    b. It must be equal to $A + B$

    c. It must be larger than A

    d. It must be positive

    Justify your choice:

2. Three vectors $\vec{a}, \vec{b},$ and $\vec{c}$ have equal magnitudes but different directions: $\vec{a}$ is directed at 30° W of S, $\vec{b}$ is directed at 30° N of W, and $\vec{c}$ is directed at 60° S of E. Using a coordinate system in which $+x$ points east and $+y$ points north, which of the following correctly ranks the x-component magnitudes of the three vectors?

    a. $c_x > a_x > b_x$

    b. $a_x = b_x > c_x$

    c. $b_x > a_x = c_x$

    d. $b_x > c_x > a_x$

    Justify your choice:

For the next two questions: An object is thrown from a height of 1 meter above the ground. At time t_0 it is released with speed v_0 at an angle of 60° above the horizontal. The object moves as a projectile reaching its highest point at time t_1, then falls down until it is again 1 meter above the ground at time t_2.

3. When was the object accelerating down at 9.8 m/s²?

    a. the entire time, from t_0 to t_2

    b. from t_0 to t_2, but momentarily not at t_1

    c. only from t_0 to t_1

    d. only from t_1 to t_2

    Justify your choice:

4. When did the object have a speed of $0.5v_0$?

    a. only once, at a time between t_0 and t_1

    b. only once, at time t_1

    c. only once, at a time between t_1 and t_2

    d. twice: once before t_1 and once after t_1

    Justify your choice:

5. A ball is to be thrown at 5.00 m/s. Which of the following angles is nearest to the angle it could be thrown (above the horizontal) so that it can be caught 5.00 m away at the same height from which it was thrown?

    a. 15°

    b. 30°

    c. 45°

    d. It can't be caught 5.00 m away in this way.

    Justify your choice:

48   Chapter 3 Review

End of Chapter Problems

Solve each of the following problems on separate paper.

1. A chemistry textbook is thrown horizontally at 16.0 m/s towards a wall that is 4.00 meters away. How much does the textbook drop before it hits the wall?

2. A biology textbook is thrown horizontally at 7.00 m/s. It is released from a height of 1.50 m above the floor. Find how far away (horizontally) the book lands on the floor.

3. You briefly consider throwing your copy of *Strive for a 5: Preparing for the AP® Physics 1 Examination*. You then decide that you won't throw it and instead place it lovingly on your bookshelf so that it will be kept safe and be ready for you whenever you want to read it. Did you make the right decision?

4. A soccer ball initially on the ground is kicked, resulting in it leaving the ground at 50.0 km/h, initially moving in a direction of 30.0° above the horizontal. What is the ball's speed 1.00 second after it was kicked?

5. A ball is on the ground. It is kicked, giving it an initial speed of 8.00 m/s. The ball leaves the ground at an angle of 20.0° above the horizontal, flies through the air, and lands back on the ground. What is the magnitude of its displacement?

Chapter 4
Forces and Motion I: Newton's Laws

Before you Read the Chapter: Prepare Yourself

Be sure that you have a good understanding of:

- Working with vectors (especially adding vectors) from **Chapter 3**
- The object and system models from **Chapter 1**

Chapter Overview

While it is true that every chapter is important, **this chapter is seriously, no kidding, for real, SUPER IMPORTANT**. In particular, this chapter introduces **Forces and Newton's three laws of motion**. These laws are easy to "remember," but it can be difficult to understand them deeply enough to consistently apply them correctly when answering questions and solving problems.

Chapter 4 is also loaded with new and important vocabulary and equations. Fortunately, Chapter 5 (which builds directly on Chapter 4) will be relatively short and (in my opinion) easier—but only if you gain a good understanding of the foundational material introduced here in Chapter 4.

Learning Objectives

☐ Describe a force as an interaction between two objects or systems and identify both objects or systems for any force.

☐ Analyze a scenario and make claims about the forces exerted on an object by other objects or systems for different types of interactions.

☐ Use Newton's second law to predict the motion of an object subject to forces exerted by other objects in a variety of physical situations.

☐ Categorize forces as long-range or contact forces, and make claims about contact forces due to the microscopic cause of those forces.

☐ Recognize the distinctions among mass, weight, and inertia.

☐ Draw and use free-body diagrams in problems to analyze situations involving multiple forces exerted on an object.

☐ Construct explanations of physical situations involving the interaction of objects using Newton's third law and the representation of action–reaction pairs of forces.

☐ Use a free-body diagram and Newton's second law to construct a mathematical representation relating the acceleration of an object to its mass, and to solve that equation for an unknown quantity.

4-1 How objects move is determined by their interactions with other objects, which can be described by forces

The foundational idea of **force** is introduced in this section. From this point onward, force will be one of the most important concepts that will be used nearly continuously. Before moving on to the next section, be sure that you understand what force is, and the difference between the terms "force," and "net external force."

While you Read the Section: Important Terms and Equations

Use the space below to define each term in your own words. You may also add any other notes that will be helpful for future review.

Force

kinetic friction

net external force (net force)

After you Read the Section: Check Your Understanding

Choose the best answer to each of the following. Use the space provided to write a short justification for your selection. When you're finished, check that you got the right answers for the right reasons!

1. Which one of the following is the best description of what force is?

    a. a push

    b. a pull

    c. a push or a pull

    d. an interaction between two objects or systems

    Justify your choice:

2. Which one of the following best determines how an object moves?

    a. whether or not there are any forces exerted on it

    b. whether or not it is interacting with anything else

    c. the combined effect of all the forces exerted on it

    d. how many forces are exerted on it

    Justify your choice:

3. If one object exerts force on a second object, then which of the following must also happen?

    a. The second object must exert a force onto the first object

    b. The first object must not move

    c. The first object must move

    d. The second object must move

    Justify your choice:

4. If an object is described as having "no net force" then what must be true about the object?

    a. there are no forces exerted on the object, or each of the forces exerted on the object must have a magnitude of zero

    b. there are either no forces exerted on the object, or all forces exerted on the object add to zero

    c. there must be no forces exerted on the object

    d. the object must be stationary

    Justify your choice:

4-2 If a net external force is exerted on an object, the object accelerates

There are **16 new terms in this section, and every single one of them is extremely important**. Try to not get overwhelmed by this. Take the time you need to learn each term and review until you understand them. You will get to work with them in more detail, deepening your understanding based on this solid foundation, in later chapters.

This section also introduces four new equations—one of which (**Newton's second law**) gets my vote as being **the most important equation of the entire course**. Although this equation is introduced here, your understanding of it will continue to develop throughout the rest of this chapter, as well as throughout Chapter 5 and beyond. It might look like a small, simple equation, but it will take time and effort to master using it in problems.

While you Read the Section: Important Terms and Equations

Use the space below to define each term in your own words. For equations, use the space to identify what each letter represents and its associated SI unit. You may also add any other notes that will be helpful for future review.

gravitational force

contact forces

normal force

kinetic friction force

static friction force

external forces

internal forces

Newton's laws of motion

Newton's second law

Inertia

inertial mass

gravitational mass

mass

kilogram (kg)

newton (N)

pound-force (lbf)

$$\Sigma \vec{F}_{ext} = \vec{F}_1 + \vec{F}_2 + \vec{F}_3 + ...$$

$$\vec{a} = \frac{\Sigma \vec{F}_{ext}}{m}$$ *Really Important!*

After you Read the Section: Check Your Understanding

Choose the best answer to each of the following. Use the space provided to write a short justification for your selection. When you're finished, check that you got the right answers for the right reasons!

1. What are the fundamental dimensions of force? *Hint: Review Section 1-6 if you need a refresher on dimensional analysis.*

    a. $(\text{mass}) \times (\text{acceleration})$

    b. $\dfrac{(\text{mass})}{(\text{acceleration})}$

    c. $\dfrac{(\text{mass}) \times (\text{distance})}{(\text{time})}$

    d. $\dfrac{(\text{mass}) \times (\text{distance})}{(\text{time})^2}$

    Justify your choice:

2. An object has four external forces exerted on it: 5 N to the east, 4 N to the west, 3 N to the east, and 2 N to the west. What is the magnitude of the net external force on the object?

    a. 0

    b. 2 N

    c. 9 N

    d. 14 N

    Justify your choice:

3. An object moves south with constant velocity. What must be true about the forces exerted on the object?

    a. the object may have external forces exerted on it, but must have zero net external force exerted on it

    b. the object must not have any external forces exerted on it

    c. the object must have a net external force directed to the south exerted on it

    d. the object must have a net external force directed to the north exerted on it

    Justify your choice:

4. Object one has net external force ΣF_1 and object two has net external force $\Sigma F_2 > \Sigma F_1$. How do the accelerations of the two objects compare?

    a. $a_2 > a_1$

    b. $a_2 = a_1$

    c. $a_2 < a_1$

    d. The comparison of acceleration cannot be made without knowing how their masses compare.

    Justify your choice:

5. Two objects have equal nonzero net external forces exerted on them. Object two has $m_2 > m_1$. How do the accelerations of the two objects compare?

    a. $a_2 > a_1$

    b. $a_2 = a_1$

    c. $a_2 < a_1$

    d. The comparison of acceleration cannot be made without more information.

    Justify your choice:

6. Two objects have equal mass. Object two has $\Sigma F_2 > \Sigma F_1$. How do the accelerations of the two objects compare?

    a. $a_2 > a_1$

    b. $a_2 = a_1$

    c. $a_2 < a_1$

    d. The comparison of acceleration cannot be made without more information.

    Justify your choice:

4-3 Mass and weight are distinct but related concepts

There are a number of words that have precise meanings in physics that differ from how the words are generally used outside of physics. This section gives definitions for two such words (mass and weight). These two words are used interchangeably outside of physics, but they have very different meanings for us, so be sure to use these words correctly!

While you Read the Section: Important Terms and Equations

Use the space below to define each term in your own words. For equations, use the space to identify what each letter represents and its associated SI unit. You may also add any other notes that will be helpful for future review.

weight

vacuum

Newton's first law

equilibrium

$F_g = mg$

If $\Sigma \vec{F}_{ext} = 0$, then $\vec{a} = 0$ and $\vec{v} = $ constant

After you Read the Section: Check Your Understanding

Choose the best answer to each of the following. Use the space provided to write a short justification for your selection. When you're finished, check that you got the right answers for the right reasons!

1. Which one of the following is a *possible* mass of an object?
   a. 5 N
   b. 10 lbf
   c. 15 kg
   d. All of the above

   Justify your choice:

2. Which one of the following is a *possible* weight of an object?
   a. 2 lbf
   b. 4 N
   c. 6 mN
   d. All of the above

   Justify your choice:

3. What is the weight of a 5 kg object that is near the surface of Earth?
   a. 5 kg
   b. 5 N
   c. 50 kg
   d. 50 N

   Justify your choice:

4. What is the acceleration due to gravity on a planet where a 160 kg object weighs 80 N?
   a. 0.5 m/s^2
   b. 2 m/s^2
   c. 9.8 m/s^2
   d. 80 m/s^2

   Justify your choice:

4-4 A free-body diagram is essential in solving any problem involving forces

This is a rather short section that introduces two new terms and no new equations. We are getting only a preview of center of mass here—much more will be covered on that topic in Chapter 9. Free-body diagrams can be helpful when working on force problems. Note that these are not "just diagrams" as **there are rules for drawing free-body diagrams, and these rules are strictly enforced on the AP® Physics 1 exam.** It will be best for you to learn these rules now and consistently use them so that you develop good habits. The **Exam Tip box on Page 164** summarizes these rules well. It would be helpful for you to **summarize these rules for future review in the space provided below.**

While you Read the Section: Important Terms and Equations

Use the space below to define each term in your own words. For equations, use the space to identify what each letter represents and its associated SI unit. You may also add any other notes that will be helpful for future review.

free-body diagram

center of mass

Rules for Free-Body Diagrams:

After you Read the Section: Check Your Understanding

Choose the best answer to each of the following. Use the space provided to write a short justification for your selection. When you're finished, check that you got the right answers for the right reasons!

1. Which one of the following statements about the center of mass of a system is true?

    a. A system can have two or more centers of mass if objects in the system have different velocities.

    b. The system moves as if all of its mass was located at its center of mass.

    c. The center of mass of a system is always located at the exact center of the system.

    d. The center of mass of a system is never located at the exact center of the system.

    Justify your choice:

2. Which one of the following statements about the object model is true?

    a. The object model can be used only for objects that have negligible size and so can be reasonably treated as idealized points.

    b. The object model can be used only for systems that are perfectly symmetrical and so can reasonably be treated as if all of its mass was at its center.

    c. The object model can be used only for systems that are composed of one piece of matter.

    d. To use the object model you need to identify the center of mass of the system.

    Justify your choice:

For the next 5 questions: A student draws the free-body diagram shown for each scenario described in the question. Does the student's drawing correctly follow all of the rules for drawing free-body diagrams on the AP® Physics 1 exam?

If not, identify all (there may be more than one) of the rules that it breaks by selecting from a, b, and c below. Select d only if the diagram follows all of the rules.

a. No. All force vectors do not originate on and touch the dot representing the object.

b. No. Vector components should never be drawn on free-body diagrams.

c. No. Not all vectors are forces that are labelled clearly with what kind of force it is.

d. Yes. This diagram follows all of the rules for free-body diagrams on the AP® Physics 1 exam.

3. An object is falling freely with only the force of gravity being exerted on it.
   Justify your choice:

4. An object is falling freely with only the force of gravity being exerted on it.
   Justify your choice:

5. An object is falling freely with only the force of gravity being exerted on it.
   Justify your choice:

6. An object has only one force exerted on it. This force has +x and +y components.
   Justify your choice:

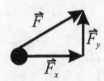

7. An object has two external forces exerted on it: one force is directed to the north and one force is directed to the north. The net external force is northeast.
   Justify your choice:

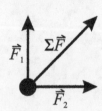

4-5 Newton's third law relates the forces that two objects exert on each other

Although Newton's third law is relatively easy to "memorize," it is frequently misunderstood. Questions on the AP® Physics 1 exam typically check for a deeper understanding, so be sure to read this section carefully.

While you Read the Section: Important Terms and Equations

Use the space below to define each term in your own words. For equations, use the space to identify what each letter represents and its associated SI unit. You may also add any other notes that will be helpful for future review.

Newton's third law

force pair

reaction force

$\vec{F}_{\text{A on B}} = -\vec{F}_{\text{B on A}}$

After you Read the Section: Check Your Understanding

Choose the best answer to each of the following. Use the space provided to write a short justification for your selection. When you're finished, check that you got the right answers for the right reasons!

1. An object has two external forces exerted on it: a downward gravitational force and an upward normal force. Which one of the following statements about these two forces is correct?

    a. These two forces form a force pair described by Newton's third law; they must be equal in magnitude and opposite in direction.

    b. These two forces do not form a force pair described by Newton's third law but must be equal in magnitude and opposite in direction.

    c. These two forces do not form a force pair described by Newton's third law; they may or may not be equal in magnitude and opposite in direction.

    d. These two forces do not form a force pair described by Newton's third law; they may not be equal in magnitude and opposite in direction.

    Justify your choice:

2. Earth exerts a downward gravitational force on a falling object. Which one of the following is the reaction force to this force?

    a. The object exerts an upward gravitational force on Earth.

    b. The air exerts an upward force on the object as it falls.

    c. The ground exerts an upward force on the object when it lands on the ground.

    d. There is no reaction force to this force.

    Justify your choice:

3. Two people each hold one end of a light rope. One person pulls their end of the rope with a force to the right and the other person pulls the other end of the rope with a force to the left. Each end of the rope pulls back with a tension force. Provided that the mass of the rope can be ignored, how do the magnitudes of the tension exerted at the two ends of the rope compare?

    a. The magnitudes of the tension at the two ends of the rope will be equal in magnitude to the force exerted by the person at that end; these magnitudes may or may not be equal.

    b. The magnitudes of the tension at the two ends of the rope will equal if the rope does not move but will be unequal if the rope moves.

    c. The magnitudes of the tension at the two ends of the rope will necessarily be unequal to each other.

    d. The magnitudes of the tension at the two ends of the rope will necessarily be equal to each other.

    Justify your choice:

4-6 All problems involving forces can be solved using the same series of steps

This section wraps up the chapter by introducing a few new important ideas and then applying them and the material introduced earlier to a wide range of example problems. These examples are meant to help you learn how to solve problems that are based on force and acceleration. While looking over the examples it is important that you do not merely attempt to "memorize the steps," but rather understand how the problem is solved so that you will be able to solve a wide range of different problems.

While you Read the Section: Important Terms and Equations

Use the space below to define each term in your own words. For equations, use the space to identify what each letter represents and its associated SI unit. You may also add any other notes that will be helpful for future review.

frame of reference

inertial frame of reference

apparent weight (effective weight)

$\Sigma F_{ext,x} = ma_x$

$\Sigma F_{ext,y} = ma_y$

After you Read the Section: Check Your Understanding

Choose the best answer to each of the following. Use the space provided to write a short justification for your selection. When you're finished, check that you got the right answers for the right reasons!

1. When solving a problem that involves forces on an object, which of the following best describe the coordinate axes x and y?

    a. The coordinate axis x must be horizontal, and y must be vertical.

    b. One of the coordinate axes x or y must be along the direction of the object's velocity.

    c. One of the coordinate axes x or y must be along the direction of the object's acceleration.

    d. The coordinate axes must be perpendicular to each other but can be oriented in any direction; aligning them with forces or acceleration is not required but is often helpful.

    Justify your choice:

2. Which one of the following best describes the steps taken to solve most problems involving force?

    a. Select the equation needed to solve the problem from the list of equations, then substitute known values into the equation and solve for the unknown.

    b. Draw one or more free-body diagrams, then write equations based on Newton's second law for each coordinate axis for each object and solve for the unknowns.

    c. Draw a single diagram showing all aspects of what is in the problem, then reflect on the situation until the answer occurs to you.

    d. Write each one of Newton's laws for each object in the problem, then solve for the unknowns.

    Justify your choice:

After you Read the Chapter: Test Yourself

After reading the chapter and trying the questions below, I recommend that you then **work on the Chapter 4 Review Problems** at the end of the chapter in your textbook before continuing into the next chapter.

Word Match

A) Center of Mass
B) Contact Forces
C) Equilibrium
D) Force
E) Free-body Diagram
F) Friction Force
G) Gravitational Force
H) Gravitational Mass
I) Inertia
J) Inertial Mass
K) Internal Forces
L) Kilogram
M) Net Force
N) Newton
O) Newton's 1st Law
P) Newton's 2nd Law
Q) Newton's 3rd Law
R) Normal Force
S) Reaction Force
T) Tension Force
U) Weight

___ 1. The type of mass pertaining to how strongly an object resists acceleration.

___ 2. The condition of having no net force, and so no acceleration.

___ 3. Description of two objects interacting with each other by touch.

___ 4. The relationship between acceleration, mass, and net force.

___ 5. The "law of inertia" – objects with no net force exerted on them do not accelerate.

___ 6. The strength of attraction to a nearby planet such as Earth.

___ 7. An object moves as if all of its mass were located at this point.

___ 8. Property of an object related to how strongly it gravitationally interacts with other objects.

___ 9. A force that one surface exerts on another, perpendicular to the surfaces.

___ 10. A force that opposes one surface from sliding across another surface.

___ 11. The SI unit of force.

___ 12. An interaction between two objects or systems.

___ 13. When two objects interact, both exert equal and opposite forces onto the other.

___ 14. Interactions between parts of a system which cannot affect the acceleration of the system.

___ 15. A downward pull from Earth.

___ 16. The sum of all external forces exerted on an object or a system.

___ 17. The SI unit of mass.

___ 18. Term sometimes used for the "other force" referenced by Newton's 3rd law.

___ 19. A property of matter related to its tendency to maintain a constant velocity.

___ 20. A tool for visualizing the forces that are exerted onto an object.

___ 21. A force provided by a string or rope.

End of Chapter Multiple Choice Questions

1. An object falls downwards with constant velocity due to the presence of air resistance (an upward drag force exerted onto the object from the air). Which of the following is the reaction force to this drag force?

   a. the downward force of gravity exerted by Earth on the object

   b. the downward force exerted by the object on the air

   c. the upward force of gravity exerted by the object on Earth

   d. the upward force exerted by Earth on the air

2. An airplane is being flown. If the airplane and everything inside of it is modeled as a single object, which of the following forces would be an internal force?

   a. the force of gravity Earth exerts on the plane

   b. the force of gravity Earth exerts on the pilot

   c. the force the air exerts on the wings

   d. the force a passenger exerts on their seat

3. An object with a mass of 5 kg is initially at rest on the floor. A vertical string with negligible mass is then attached to the object and the other end of the string has an upward force of 30 N exerted on it. Which of the following is the magnitude of the normal force that would then be exerted on the object by the floor? (Use $g = 10.0$ N/kg)

   a. 50 N

   b. 30 N

   c. 20 N

   d. 0 N (the object loses contact with the floor)

4. An object moving in the $+x$ direction has three forces exerted on it as seen in the free-body diagram for the object shown here. Force $\vec{F_1}$ points in the $+x$ direction and force $\vec{F_2}$ points in the $-y$ direction. Which of the following is needed for the *magnitudes* of the forces in order for the object to continue to move in the $+x$ direction with constant velocity?

   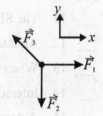

   a. $|\vec{F_3}| = \sqrt{|\vec{F_1}|^2 + |\vec{F_2}|^2}$

   b. $|\vec{F_3}| = |\vec{F_1}| + |\vec{F_2}|$

   c. $|\vec{F_3}| < |\vec{F_1}|$

   d. $|\vec{F_3}| < \sqrt{|\vec{F_1}|^2 + |\vec{F_2}|^2}$

5. Objects with masses $m_1 = 100$ kg and $m_2 = 0.1$ kg are attached by a string of negligible mass that passes over a pulley that turns with negligible friction as shown here. There is negligible friction between m_2 and the surface it is on. Which of the following is nearest to the tension in the string?

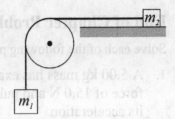

   a. 1 N
   b. 100 N
   c. 500 N
   d. 1000 N

End of Chapter Problems

Solve each of the following problems on separate paper. Use $g = 9.8$ m/s².

1. A 5.00 kg mass has exactly two forces exerted on it. It is simultaneously pushed east with a force of 15.0 N and pulled west with a force of 25.0 N. Find the magnitude and direction of its acceleration.

2. A 1200.0 kg car goes from 0 to 25.0 m/s in 8.00 seconds. What was the magnitude of the net force exerted on the car during this acceleration?

3. What is the weight of a 5.0 kg object?

4. A 2.50 kg object sits on the floor. What is the magnitude of the normal force exerted on the object?

5. A 2.5 kg object sits on the floor of an elevator which is accelerating upwards at 3.0 m/s². What is the size of the normal force on the object?

6. What is the apparent weight of a 2.5 kg object while it is in an elevator that is accelerating downward at 3.0 m/s².

7. A 1500.0 kg car is moving on the road at (initially) 20.0 m/s. The brakes are then fully applied, resulting in the car's skidding to a stop. How long are the skid marks if the force of friction exerted by the road on the car while it is skidding is 12000.0 N?

8. A surface with negligible friction is inclined at 30.0° to the horizontal. Find the magnitude of the acceleration of an object that is sliding down the incline.

For the next two problems: Two objects with mass $m_1 = 7.0$ kg and $m_2 = 5.0$ kg are connected by a string that wraps around a pulley as shown. Both objects are released from rest. Assume the string does not stretch, and that the pulley has negligible mass or friction with the axle.

9. What is the magnitude of the tension in the string?

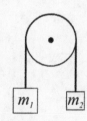

10. What is the magnitude of the acceleration of the objects?

70   Chapter 4 Review

Chapter 5
Forces and Motion II: Applications

Before you Read the Chapter: Prepare Yourself

Be sure that you have a good understanding of:

- **Chapter 4**—all of it! It is generally true that you need to understand a chapter before moving on to the next one, but it is especially true here!
- **Trigonometry**—especially the "tan" of an angle. Review this from the Math Tutorial, section M8.

Chapter Overview

Chapter 5 is a direct continuation of the material that was introduced in Chapter 4. The focus is once again force and Newton's laws, but now that the fundamentals have been covered, Chapter 4 expands on these ideas by applying them to a greater range of problems. This chapter also expands on the force of friction and introduces drag and spring forces.

Learning Objectives

- ☐ Recognize what determines the magnitude of the static friction force and find the magnitude and direction of the maximum static friction force exerted on an object by a surface.

- ☐ Find the magnitude and direction of the force of kinetic friction exerted on an object by a surface.

- ☐ Explain contact forces on an object as arising from interatomic electric interactions and determine the direction in which these forces must therefore be exerted on the object.

- ☐ Construct free-body diagrams, and extract quantitative or qualitative information from them to solve for properties of the motion of an object when the forces on an object include static or kinetic friction.

- ☐ Analyze situations in which fluid resistance is important to determine when application of constant acceleration equations would give poor results.

- ☐ Analyze equilibrium force problems involving the spring force to solve for properties of the system such as mass, spring constant, or compression or extension of the spring.

5-1 We can use Newton's laws in situations beyond those we have already studied

No new equations or important terms are introduced in this section—it is introductory, but serves as an important transition point. When learning about Physics it is usually best to initially simplify things by ignoring or limiting complicating factors such as friction and air resistance. Only after establishing a good understanding of the underlying principles are we able to then add layers of complexity.

In Chapter 4, friction was often assumed to be negligible (i.e., so small that it is safe to ignore its presence), and when it was present, it was not much different than any other external force. However, there is a lot more to learn about friction. I like to describe friction as the "trickiest" force that we will encounter in all of AP® Physics 1. There are some equations about friction that will be introduced later in the chapter, but for now, try out the following questions to make sure that you are ready to proceed!

After you Read the Section: Check Your Understanding

Choose the best answer to each of the following. Use the space provided to write a short justification for your selection. When you're finished, check that you got the right answers for the right reasons!

1. A block slides to the north on a stationary surface. In which direction is the force of friction exerted on the block by the surface?

   a. north   b. south   c. down

   d. There is no friction force exerted onto the object by the surface.

   Justify your choice:

2. The surface of a conveyor belt moves to the north. A block is placed on the surface and prevented from moving by an external force. In which direction is the force of friction exerted on the block by the surface?

   a. north   b. south   c. down

   d. There is no friction force exerted onto the object by the surface.

   Justify your choice:

3. Which one of the following statements about kinetic friction is true?

   a. The kinetic friction force is exerted on an object only if the object is moving.

   b. The kinetic friction force is exerted on an object only if the object is stationary.

   c. The kinetic friction force is exerted on an object only if the object is moving relative to a surface that it is in contact with.

   d. The kinetic friction force is exerted on an object only if the object exerts a static friction force on a surface that it is in contact with.

   Justify your choice:

5-2 The static friction force changes magnitude to offset other forces being exerted on a system

I already mentioned that I think that friction is a "tricky" force. This section reveals how the static friction force can be especially tricky; for a pair of surfaces in contact with each other there is a *maximum* amount of static friction available, but the actual magnitude of this force depends on other conditions. This is why we have an equation for the maximum available static friction but must use an inequality for its magnitude more generally.

Another reason that the friction force is tricky is that the normal force is often needed in order to calculate friction forces. The normal force is somewhat tricky on its own! Be sure to read the **Watch Out!** and **AP® Exam Tip** boxes on the bottom of Page 200, as they explain one of the most common misconceptions about the normal force.

While you Read the Section: Important Terms and Equations

Use the space below to define each term in your own words. For equations, use the space to identify what each letter represents and its associated SI unit. You may also add any other notes that will be helpful for future review.

macroscopic

coefficient of static friction μ_S

$F_S \leq F_{S,\text{max}}$

$F_{S,\text{max}} = \mu_S F_n$

$\mu_S = \tan \theta_{\text{slip}}$

After you Read the Section: Check Your Understanding

Choose the best answer to each of the following. Use the space provided to write a short justification for your selection. When you're finished, check that you got the right answers for the right reasons!

1. A 2.00 kg block is at rest on a level surface. The coefficient of static friction between the block and the surface is 0.500. What is the magnitude of the static friction force exerted onto the block by the surface?

   a. 0   b. 1.00 N   c. 5.00 N   d. 10.0 N

   Justify your choice:

2. A 2.00 kg block is at rest on a level surface. The coefficient of static friction between the block and the surface is 0.500. An external force of 5.00 N to the east is then exerted on the block. What is the magnitude of the static friction force exerted on the block by the surface?

   a. 0   b. 1.00 N   c. 5.00 N   d. 10.0 N

   Justify your choice:

3. A 2.00 kg block is at rest on a level surface. The coefficient of static friction between the block and the surface is 0.500. An external force of 50.00 N to the east is then exerted onto the block. What is the smallest magnitude of a second external force that, if exerted onto the block in the downward direction, would prevent the block from sliding?

   a. 100.0 N   b. 80.0 N   c. 50.0 N   d. 20.0 N

   Justify your choice:

4. A block is placed on a surface that is initially level. The coefficient of static friction between the block and the surface is 1.20. One end of the surface is then slowly lifted to gradually increase how tilted the surface is. Which of the following is nearest to the angle of tilt at which the block will begin to slide?

   a. 20°

   b. 30°

   c. 40°

   d. 50°

   Justify your choice:

5-3 The kinetic friction force on a sliding object has a constant magnitude

This section shifts the focus away from static friction to kinetic friction, and also includes an introduction to rolling friction.

While you Read the Section: Important Terms and Equations

Use the space below to define each term in your own words. For equations, use the space to identify what each letter represents and its associated SI unit. You may also add any other notes that will be helpful for future review.

coefficient of kinetic friction μ_k

rolling friction

coefficient of rolling friction μ_r

$F_k = \mu_k F_n$

After you Read the Section: Check Your Understanding

Choose the best answer to each of the following. Use the space provided to write a short justification for your selection. When you're finished, check that you got the right answers for the right reasons!

1. An object slides on a surface which exerts a kinetic friction force on the object. Increasing which one of the following would necessarily increase the magnitude of the kinetic friction force?

    a. the speed of the object

    b. the acceleration of the object

    c. the magnitude of the normal force

    d. the contact area between the object and the surface

    Justify your choice:

2. What are the fundamental dimensions of the coefficient of kinetic friction?

    a. (mass)

    b. $(\text{mass}) \times (\text{length})$

    c. $\dfrac{(\text{mass}) \times (\text{length})}{(\text{time})^2}$

    d. None of the above; the coefficient of kinetic friction is dimensionless.

    Justify your choice:

3. While a bicycle is moving, the rider applies the brakes hard enough for the tires to suddenly stop turning and skid on the road. Which one of the following forces is the external force that then slows the bicycle-rider system to a stop?

    a. static friction

    b. kinetic friction

    c. rolling friction

    d. the force that the rider exerts onto the brake

    Justify your choice:

5-4 Problems involving friction are solved like any other force problems

This section wraps up the topic of friction by applying the ideas from the previous sections and the material introduced earlier to a wide range of example problems. While looking over the examples it is important that you not merely attempt to "memorize the steps," but rather reach an understanding of how the problem is solved so that you will be able to solve a wide range of different problems.

This section does not introduce any new terms or equations.

After you Read the Section: Check Your Understanding

Choose the best answer to each of the following. Use the space provided to write a short justification for your selection. When you're finished, check that you got the right answers for the right reasons!

1. A block slides across a level surface. The only horizontal force exerted on the object is the friction force exerted onto the block by the surface. If the object slows down at a rate of 5.0 m/s², then what is the coefficient of kinetic friction between the surface and the block?

   a. 0.50
   b. 2.0
   c. 5.0
   d. More information is needed to calculate the coefficient of kinetic friction.

   Justify your choice:

2. A block is on the floor of an elevator. When the elevator is at rest, an external force with magnitude F_1 directed horizontally is needed to start the block sliding. Later it is found that a force with magnitude $F_2 > F_1$ is needed to start the block sliding. Which one of the following could explain why the larger force is then needed?

   a. The elevator is moving upwards with constant speed.
   b. The elevator is moving downwards with constant speed.
   c. The elevator is moving downwards with increasing speed.
   d. The elevator is moving downwards with decreasing speed.

   Justify your choice:

3. A 2 kg block is initially at rest on a level surface. How much force must be exerted on the block by a person in order to give the block an acceleration of 4 m/s² if the coefficient of kinetic friction between the block and the surface is 0.5?

   a. 8 N
   b. 10 N
   c. 18 N
   d. 20 N

   Justify your choice:

5-5 An object moving through air or water experiences a drag force

In AP® Physics we generally assume that the motion of objects is not affected by the presence of air. This is often described as there being "no air resistance" or "negligible air resistance." With that said, there are situations in which on an object moving through the air (or water) has a drag force exerted on it that cannot be neglected. This section introduces some terminology and equations on this topic.

While you Read the Section: Important Terms and Equations

Use the space below to define each term in your own words. For equations, use the space to identify what each letter represents and its associated SI unit. You may also add any other notes that will be helpful for future review.

fluid resistance

drag force

$F_{drag} = bv$

$F_{drag} = cv^2$

After you Read the Section: Check Your Understanding

Choose the best answer to each of the following. Use the space provided to write a short justification for your selection. When you're finished, check that you got the right answers for the right reasons!

1. While moving through a fluid at low speed, a small object has a drag force with magnitude F_1 exerted on it by the fluid. If the speed of the object is halved, what will be the magnitude of the drag force?

    a. $0.5 F_1$

    b. F_1

    c. $2 F_1$

    d. $4 F_1$

    Justify your choice:

2. While moving through a fluid at high speed, a large object has a drag force with magnitude F_1 exerted on it by the fluid. If the speed of the object is doubled, what will be the magnitude of the drag force?

    a. $0.5 F_1$

    b. F_1

    c. $2 F_1$

    d. $4 F_1$

    Justify your choice:

3. Which one of the following correctly describes the directions of a person's velocity and acceleration immediately after opening their parachute, if that person is then falling downward with a speed greater than their terminal speed?

    a. velocity is upward, acceleration is downward

    b. velocity is downward, acceleration is upward

    c. velocity and acceleration are both downward

    d. velocity and acceleration are both upward

    Justify your choice:

5-6 An ideal spring force can be used to model many interactions

This is a rather short section in which springs are introduced. As Chapter 5 is a chapter on Forces and Motion, the fact that springs can exert forces is emphasized here; other aspects of springs will come in a number of future chapters, so it is worth spending time now to reach a good understanding of spring forces.

While you Read the Section: Important Terms and Equations

Use the space below to define each term in your own words. For equations, use the space to identify what each letter represents and its associated SI unit. You may also add any other notes that will be helpful for future review.

ideal spring

Hooke's law

spring constant

$F_{s,x} = -k\Delta x$

After you Read the Section: Check Your Understanding

Choose the best answer to each of the following. Use the space provided to write a short justification for your selection. When you're finished, check that you got the right answers for the right reasons!

1. What are the fundamental dimensions of the spring constant?

   a. $\dfrac{(\text{mass}) \times (\text{length})}{(\text{time})^2}$
   b. $\dfrac{(\text{time})^2}{(\text{mass}) \times (\text{length})}$
   c. $\dfrac{(\text{mass})}{(\text{time})^2}$
   d. $\dfrac{(\text{length})}{(\text{time})^2}$

   Justify your choice:

2. A spring is 30 cm long when it is relaxed. With the upper end of the spring fixed in position, a 2 kg object is attached to the lower end of the spring. The object is then slowly lowered down until the spring is 50 cm long, at which point the spring fully supports the object hanging from its lower end, holding it at rest. What is the spring constant of this spring?

   a. 1 N/m
   b. 10 N/m
   c. 40 N/m
   d. 100 N/m

   Justify your choice:

3. A student measures the magnitude of the force F_s exerted by a spring when it is stretched by an amount Δx for a range of values for Δx. How can this data be used to graphically determine the spring constant of the spring?

   a. Plot F_s on the vertical axis and Δx on the horizontal axis; the spring constant is equal to the slope of this graph.
   b. Plot F_s on the vertical axis and Δx on the horizontal axis; the spring constant is equal to the area of this graph.
   c. Plot Δx on the vertical axis and F_s on the horizontal axis; the spring constant is equal to the slope of this graph.
   d. Plot Δx on the vertical axis and F_s on the horizontal axis; the spring constant is equal to the slope of this graph.

   Justify your choice:

After you Read the Chapter: Test Yourself

After reading the chapter and trying the questions below, I recommend that you then **work on the Chapter 5 Review Problems** at the end of the chapter in your textbook before continuing into the next chapter.

End of Chapter Multiple Choice Questions

1. An object with mass m is placed on a level surface. A person then exerts a force with magnitude $5F$ on the object. The object slides with a force of kinetic friction with magnitude $3F$ also exerted on it by the surface. Which of the following is a correct expression for the object's acceleration?

    a. $\dfrac{F}{m}$
    b. $\dfrac{2F}{m}$
    c. $\dfrac{4F}{m}$
    d. $\dfrac{8F}{m}$

2. A block with mass m is placed on top of a larger block with mass M. The coefficient of static friction between the two blocks is μ_S, but friction between the larger block on the surface that supports it is negligible. A force of magnitude F is then exerted onto the larger block, causing both blocks to accelerate together to the right, as shown here. Which of the following is an expression for the acceleration of the blocks?

    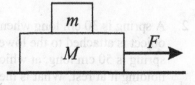

    a. $\dfrac{F}{m+M}$
    b. $\dfrac{F-\mu_S mg}{m}$
    c. $\dfrac{F-\mu_S mg}{m+M}$
    d. $\dfrac{F-\mu_S(m+M)g}{m+M}$

3. An object is dropped from rest from a large height. While falling, the object has a drag force exerted on it by the air. Which of the following best describe the magnitude of the drag force?

    a. It is zero and constant.
    b. It is nonzero and constant.
    c. It is initially small and increases with time.
    d. It is initially large and decreases with time.

4. An object is dropped from rest from a large height. While falling, the object has a drag force exerted on it by the air. Which of the following best describe the magnitude of the object's acceleration?

    a. It is zero and constant.
    b. It is nonzero and constant.
    c. It is initially small and increases with time.
    d. It is initially large and decreases with time.

5. An object is projected upward from the ground at time t_1. It reaches its maximum height at time t_2, then falls down, hitting the ground at time t_3. The object interacts with the air, and so has a drag force exerted on it by the air. How does the time that the object rises toward its maximum height ($\Delta t_{rise} = t_2 - t_1$) compare to the time it falls to the ground from its maximum height ($\Delta t_{fall} = t_3 - t_2$)?

    a. $\Delta t_{fall} < \Delta t_{rise}$
    b. $\Delta t_{fall} > \Delta t_{rise}$
    c. $\Delta t_{fall} = \Delta t_{rise}$
    d. The comparison cannot be determined without more information.

End of Chapter Problems

Solve each of the following problems on separate paper.

1. A 2.00 kg block is at rest on a table. The coefficients of friction are $\mu_k = 0.400$ and $\mu_s = 0.800$. A horizontal force to the right is then exerted on the block. Find the acceleration of the block if the applied force has a magnitude of 10.0 N.

2. A 2.00 kg block is at rest on a table. The coefficients of friction are $\mu_k = 0.400$ and $\mu_s = 0.800$. A horizontal force to the right is then exerted on the block. Find the acceleration of the block if the applied force has a magnitude of 15.0 N.

3. A 2.00 kg block is at rest on a table. The coefficients of friction are $\mu_k = 0.400$ and $\mu_s = 0.800$. A horizontal force to the right is then exerted on the block. Find the magnitude of the acceleration of the block if the applied force has a magnitude of 20.0 N.

4. An 8.00 kg block is being dragged on the ground by a rope that makes a 25.0° angle to the horizontal. The tension in the rope is 30.0 N and both coefficients of friction are equal to 0.200. Find the acceleration of the block.

5. A surface is inclined at 30.0° to the horizontal. An object is placed onto this surface and released from rest. Find the magnitude of its acceleration given that the coefficients of friction between the block and the surface are $\mu_s = 0.700$ and $\mu_k = 0.500$.

6. A surface is inclined at 30.0° to the horizontal. An object is placed onto this surface and then sent sliding down the incline by an external force that was briefly exerted onto the object by a person. Find the magnitude of the acceleration of the object while it slides down the incline after the external force exerted by the person has been removed, given that the coefficients of friction between the block and the surface are $\mu_s = 0.700$ and $\mu_k = 0.500$.

Chapter 6
Circular Motion and Gravitation

Before you Read the Chapter: Prepare Yourself

Be sure that you have a good understanding of:

- **Chapters 4 and 5**—Chapter 6 makes regular use of the ideas introduced in the last two chapters.

- **Inverse proportions** and **constants of proportionality**—review M-3 from the math tutorial.

Chapter Overview

Compared to all previous chapters, Chapter 6 introduces the fewest new equations and terms. In a way, Chapter 6 is an extension of some of the fundamental ideas already introduced. **Circular motion** is a particularly interesting kind of motion that will expand your understanding of acceleration. It is easy to add what you learned in the last two chapters to make sense of how forces come into play for an object moving in circular motion. Chapter 6 also explores **gravitation** in considerably more depth.

This is a relatively straightforward chapter, but only if (1) you reached a good understanding of the last two chapters, and (2) you are willing to rethink your understanding of acceleration. Most people's intuition about acceleration is limited. The physics introduced in this chapter makes perfect sense in the end, but it may be a struggle for you to reach this understanding. Keep at it! Reaching an understanding of this material is rewarding and worth the effort.

Learning Objectives

☐ Describe why an object moving in a circle is always accelerating even when its speed is not changing.

☐ Apply Newton's laws to objects in uniform circular motion.

☐ Recognize what it means to say that gravitation is universal, and articulate when the gravitational force is the dominant force between objects or systems.

☐ Apply Newton's law of universal gravitation to describe or calculate the gravitational force any two

☐ objects exert on each other, and use that force in contexts other than orbital motion.

☐ Apply the law of universal gravitation to analyze circular orbits of satellites and planets.

☐ Relate the gravitational field at a point in space to the gravitational force exerted on an object at that point in space.

☐ Explain the origin of apparent weightlessness.

6-1 Gravitation is a force of universal importance; add circular motion and you start explaining the motion of the planets

Although it is largely introductory, this section reveals that gravity is not merely a special feature of Earth, but rather is truly universal. We will see equations that will add details later, but take a moment after reading this section to appreciate how profound this idea is!

While you Read the Section: Important Terms and Equations

Use the space below to define each term in your own words. For equations, use the space to identify what each letter represents and its associated SI unit. You may also add any other notes that will be helpful for future review.

law of universal gravitation

After you Read the Section: Check Your Understanding

Choose the best answer to each of the following. Use the space provided to write a short justification for your selection. When you're finished, check that you got the right answers for the right reasons!

1. Which of the following pairs of objects exert gravitational forces on each other?

    a. Earth and a person on Earth

    b. the Sun and Earth

    c. a speck of dust on the Moon and a piece of paper on Earth

    d. all of the above

   Justify your choice:

2. Which one of the following can be best understood by combining universal gravity and circular motion?

    a. How an object falls down when it is dropped.

    b. How the Moon orbits around Earth.

    c. How a car can be steered around a circular path.

    d. How a ball moves as a projectile when thrown.

   Justify your choice:

6-2 An object moving in a circle is accelerating even if its speed is constant

Both this section and the next section focus on uniform circular motion—the motion of an object around a circular path with constant speed. This section introduces the fact that an object in uniform circular motion is accelerating, despite the fact that its speed remains constant. Some students have a hard time accepting this fact due to the way the word "acceleration" is often casually used to indicate that an object is specifically speeding up.

It is important to read this section carefully and come to understand how centripetal acceleration is a "real acceleration" and not merely a technicality of how physics defines acceleration. This will be especially important when we consider other aspects of uniform circular motion later in the chapter.

While you Read the Section: Important Terms and Equations

Use the space below to define each term in your own words. For equations, use the space to identify what each letter represents and its associated SI unit. You may also add any other notes that will be helpful for future review.

uniform circular motion

Period

centripetal acceleration

$$a_{cent} = \frac{v^2}{r}$$

86   Section 6-2

After you Read the Section: Check Your Understanding

Choose the best answer to each of the following. Use the space provided to write a short justification for your selection. When you're finished, check that you got the right answers for the right reasons!

1. Which one of the following objects has an acceleration?
   a. An object moving in a straight path while slowing down.
   b. An object moving with constant speed along a curved path.
   c. Both of the above.
   d. None of the above.

   Justify your choice:

2. Which one of the following statements about circular motion is true?
   a. An object moving in a circular path with constant speed is not accelerating.
   b. An object moving in a circular path with constant speed accelerates in the direction of its velocity.
   c. An object moving in a circular path with constant speed accelerates away from the center of the circle.
   d. An object moving in a circular path with constant speed accelerates toward the center of the circle.

   Justify your choice:

3. An object has acceleration a_0 when moving at speed v_0 around a circular path with radius r_0. What is the acceleration of a different object moving at speed $2v_0$ around a circular path with radius $2r_0$?
   a. $0.5a_0$
   b. a_0
   c. $2a_0$
   d. $4a_0$

   Justify your choice:

6-3 For an object in uniform circular motion, the net force exerted on the object points toward the center of the circle

This section shifts the focus to the force aspect of uniform circular motion by applying Newton's laws—especially Newton's second law—to objects moving in uniform circular motion. Although the one equation introduced in this section initially looks unfamiliar, it is important to understand that it is nothing more than Newton's second law written for the special case of uniform circular motion.

It is also important to note that unlike the normal force, gravitational force, and force of friction, the **"centripetal force" is not another specific type of force, but rather a term that can be used for the *net force* exerted on an object moving in uniform circular motion.** *Really Important!*

While you Read the Section: Important Terms and Equations

Use the space below to define each term in your own words. For equations, use the space to identify what each letter represents and its associated SI unit. You may also add any other notes that will be helpful for future review.

centripetal force

$$\Sigma F = F_{cent} = \frac{mv^2}{r}$$

After you Read the Section: Check Your Understanding

Choose the best answer to each of the following. Use the space provided to write a short justification for your selection. When you're finished, check that you got the right answers for the right reasons!

1. Which one of the following correctly describes the net force exerted on an object moving in uniform circular motion?

    a. The net force is directed toward the center of the circle.

    b. The net force is directed away from the center of the circle.

    c. The net force is directed tangent to the circle.

    d. There is no net force.

    Justify your choice:

2. An object moving in uniform circular motion has two external forces exerted on it. Which of the following must be true about these two forces?

    a. The two forces must be equal in magnitude and opposite in direction.

    b. The vector sum of the two forces must be directed toward the center of the circle.

    c. At least one of the forces must be directed toward the center of the circle.

    d. Both of the forces must be directed toward the center of the circle.

    Justify your choice:

3. Which one of the following statements about net force and centripetal force is correct?

    a. An object moving in uniform circular motion must have a centripetal force directed *toward* the center of the circle; the net force is the sum of the centripetal force and any other forces being exerted on the object.

    b. An object moving in uniform circular motion must have a centripetal force directed *away from* the center of the circle; the net force is the sum of the centripetal force and any other forces being exerted on the object.

    c. For an object moving in uniform circular motion, the net force may be called centripetal force because it must be directed *away from* the center of the circle.

    d. For an object moving in uniform circular motion, the net force may be called centripetal force because it must be directed *toward* the center of the circle.

    Justify your choice:

6-4 **Newton's law of universal gravitation explains the orbit of the Moon, and introduces us to the concept of field**

The last two sections of the chapter have been on the topic of uniform circular motion. It's important to note that while this section steps away from that topic to introduce a deeper understanding of gravity, these two separate topics will be brought together in the next section!

While you Read the Section: Important Terms and Equations

Use the space below to define each term in your own words. For equations, use the space to identify what each letter represents and its associated SI unit. You may also add any other notes that will be helpful for future review.

gravitational constant

Cavendish experiment

Field

$$F_{1 \text{ on } 2} = F_{2 \text{ on } 1} = \frac{Gm_1 m_2}{r^2}$$

$$g = \frac{Gm_{\text{Earth}}}{R_{\text{Earth}}^2}$$

After you Read the Section: Check Your Understanding

Choose the best answer to each of the following. Use the space provided to write a short justification for your selection. When you're finished, check that you got the right answers for the right reasons!

1. Two objects with masses m_0 and $2m_0$ exert gravitational forces on each other. If the less massive object exerts a force with magnitude F_0 on the more massive object, then what is the magnitude of force that the more massive object exerts on the less massive object?

    a. $0.5F_0$ b. F_0 c. $2F_0$ d. $4F_0$

    Justify your choice:

2. A star exerts a gravitational force with magnitude of F_0 onto an object when the object is at a distance r_0 from the star. If the object is moved to be a distance of $0.5r_0$ from the star, then what is the magnitude of force that the star will exert onto the object?

    a. $0.5F_0$ b. F_0 c. $2F_0$ d. $4F_0$

    Justify your choice:

3. An object with mass m is elevated to a height h above the surface of a planet with mass M and radius R. Which of the following is an expression for the weight of the object?

    a. $\dfrac{GmM}{R^2}$

    b. $\dfrac{GmM}{h^2}$

    c. $\dfrac{GmM}{(R+h)^2}$

    d. $\dfrac{GmM}{R^2+h^2}$

    Justify your choice:

6-5 Newton's law of universal gravitation begins to explain the orbits of planets and satellites

This section unites the two separate topics (uniform circular motion and universal gravitation) recently introduced. Planets and satellites in circular orbits move in uniform circular motion due to the gravitational force exerted on them by the body they orbit!

While you Read the Section: Important Terms and Equations

Use the space below to define each term in your own words. For equations, use the space to identify what each letter represents and its associated SI unit. You may also add any other notes that will be helpful for future review.

orbital period

$$v = \sqrt{\frac{Gm_{Earth}}{r}}$$

$$T^2 = \frac{4\pi^2}{Gm_{Earth}} r^3$$

After you Read the Section: Check Your Understanding

Choose the best answer to each of the following. Use the space provided to write a short justification for your selection. When you're finished, check that you got the right answers for the right reasons!

1. Which one of the following best describes what the term orbital period means?
   a. The amount of time needed for a satellite to complete one orbit.
   b. The total amount of time that a satellite has been in orbit.
   c. How often a satellite completes orbits.
   d. The total number of orbits that a satellite has completed.

   Justify your choice:

2. Which of the following is the correct relationship between orbital speed and orbital radius?
   a. $v \propto \dfrac{1}{r}$
   b. $v \propto \dfrac{1}{\sqrt{r}}$
   c. $v \propto \dfrac{1}{r^2}$
   d. $v \propto \dfrac{1}{r^3}$

   Justify your choice:

3. Several satellites orbit around the same planet. Which one of the following calculated values would be equal for all of the satellites?
   a. $\dfrac{r}{T}$
   b. $\dfrac{r^2}{T}$
   c. $\dfrac{r^3}{T^2}$
   d. $\dfrac{r^2}{T^3}$

   Justify your choice:

6-6 Apparent weight and what it means to be "weightless"

Words like "weight" and "weightless" can be problematic; in physics, we use these terms formally in an exact way that differs from how they are casually used outside of physics. Of particular note, objects that are casually described as being "weightless" do not have zero weight! In this section you will learn the real physics behind what gives the appearance of weightlessness for things and people in orbit.

While you Read the Section: Important Terms and Equations

Use the space below to define each term in your own words. For equations, use the space to identify what each letter represents and its associated SI unit. You may also add any other notes that will be helpful for future review.

apparent weightlessness (effective weightlessness)

After you Read the Section: Check Your Understanding

Choose the best answer to each of the following. Use the space provided to write a short justification for your selection. When you're finished, check that you got the right answers for the right reasons!

1. Videos of astronauts orbiting around Earth show people and objects floating around in what is sometimes described as a weightless environment. Which one of the following best explains the physics behind this phenomenon?

    a. People and objects in orbit still have substantial weight; the appearance of weightlessness is due to them being in circular motion with the outward forces balancing the force of gravity.

    b. People and objects in orbit still have substantial weight; the appearance of weightlessness is due to them being in a state of free fall along with their immediate surroundings.

    c. There is no gravity in space, so astronauts and objects in orbit around Earth really are weightless.

    d. There is very little gravity in space, so astronauts and objects in orbit around Earth are practically weightless.

    Justify your choice:

2. Physiological challenges of apparent weightlessness include which of the following?

    a. loss of muscle mass
    b. loss of bone mass
    c. increased risk of infection
    d. all of the above

    Justify your choice:

After you Read the Chapter: Test Yourself

End of Chapter Multiple Choice Questions

1. Which of the following is a correct expression for centripetal acceleration?

   a. $\dfrac{2\pi v}{T}$ b. $\dfrac{T}{2\pi v}$ c. $\dfrac{2\pi r}{T^2}$ d. $\dfrac{T^2}{2\pi r}$

2. A newly discovered planet has mass $4m_E$ and radius $4R_E$, where m_E is the mass of Earth and R_E is the radius of Earth. The acceleration due to gravity at the surface of the planet is most nearly
   a. 80 m/s²   b. 40 m/s²   c. 2.5 m/s²   d. 1.3 m/s²

3. A washing machine spins wet clothes in horizontal circular motion at high speed in a container which has small holes in it. After being spun, the clothes contain significantly less water than they did before being spun. Which of the following is the best explanation for why this is?

   a. The water and the clothes both had forces exerted on them away from the center of the spinning motion, but only the water could fit through the small holes.

   b. The water had a force exerted on it away from the center of the spinning motion which pushed it through the holes, but the clothes had a force exerted on them toward the center of the spinning motion causing them to move in circular motion.

   c. The clothes had forces exerted on them toward the center of the spinning motion, and there was an equal and opposite reaction force exerted onto the water which pushed it out of the holes.

   d. The clothes had forces exerted on them toward the center of the spinning motion and so did some of the water spinning with the clothes; water next to a hole no longer had this force exerted on it so it moved through the hole.

4. An object is moved from the surface of Earth to the surface of the Moon along a path that keeps it between Earth and the Moon at all times. The weight of the object ΣF_g is defined as the magnitude of the net gravitational force exerted on the object from Earth and the Moon. Which of the following is a possible graph of the object's weight throughout this motion?

   a. 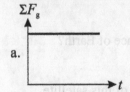   b. 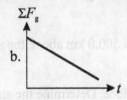   c. 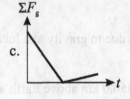   d.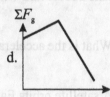

5. Star A has mass m_A and is found to have a planet in a circular orbit around it with an orbital radius r_A. Star B has a planet in a circular orbit around it with orbital radius $2r_A$. The two planets orbit their respective stars with the same orbital period. What is the mass of star B?

   a. m_A b. $2\,m_A$ c. $4\,m_A$ d. $8\,m_A$

End of Chapter Problems

Solve each of the following problems on separate paper.

1. A car moves at 50.0 km/h around a circular curve with radius 150.0 m. What is the magnitude of its acceleration?

2. Knowing that Earth orbits the Sun in a very nearly circular orbit with a radius of 1.50×10^{11} m and that it takes one year to for Earth to orbit around the Sun, calculate Earth's speed as it orbits the Sun.

3. Knowing that Earth orbits the Sun in a very nearly circular orbit with a radius of 1.50×10^{11} m and that it takes one year to for Earth to orbit around the Sun, calculate the magnitude of Earth's acceleration as it orbits the Sun.

4. A 25000 kg fighter jet moves at a constant speed of 200 m/s around a circular curve with radius 5 km. What is the centripetal force exerted on this jet?

5. A 1200.0 kg car moves around a 600.0 m radius circular racetrack, taking 1.00 minute to complete each lap. What is the centripetal force exerted on this car?

6. A 3.00 kg rock is tied to the end of a 40.0 cm long piece of string. The rock is then "twirled" around in a vertical circle. Calculate the tension in the string when the rock is at the *lowest* point of the circle, at which point its speed is 2.50 m/s.

7. A 3.00 kg rock is tied to the end of a 40.0 cm long piece of string. The rock is then "twirled" around in a vertical circle. Calculate the tension in the string when the rock is at the *highest* point of the circle, at which point its speed is 2.50 m/s.

8. A car's tires on the road have a coefficient of friction of 0.800. What is the fastest that this car can take a level turn with a radius of 50.0 m, without skidding?

9. What is the acceleration due to gravity at a location 500.0 km above the surface of Earth?

10. A satellite orbits Earth 450.0 km above Earth's surface. Determine the speed of this satellite.

11. The Sun has a mass of 2.00×10^{30} kg. If there were a planet located 1.50×10^{11} m from the Sun, what would its orbital period be?

Chapter 7
Conservation of Energy

Before you Read the Chapter: Prepare Yourself

Be sure that you have a good understanding of:

- the **object model** and **systems** (pages 8 and 9 in your textbook)
- **forces** and **free-body diagrams** (Section 4-4 in your textbook)

Memorize the following information about the cosines of angles. Yes, you could use a calculator for this, but trust me on this one... you will benefit from knowing the following without having to continuously reach for a calculator! You can use your calculator for other values when needed.

θ	$\cos(\theta)$
0°	+1
between 0° and 90°	positive (between 0 and 1)
90°	0
between 90° and 180°	negative (between 0 and -1)
180°	-1

Chapter Overview

This chapter introduces ideas related to work and energy, including the important principle of conservation. As you will see, some problems are ideally approached using one set of tools (kinematics and Newton's laws, for example) while other problems require other tools (such as work and energy).

While learning about work and energy in this and the next chapter, you may be tempted to fall back to familiar tools learned in previous chapters, especially in simple problems. Resist this temptation, as these simpler problems will help you develop skills with these new ideas that you will need in the future!

Learning Objectives

- ☐ Explain what it means for a quantity to be conserved.
- ☐ Describe what conditions must be met for work to be done on an object, for both positive and negative work.
- ☐ Explain the relationship between work and kinetic energy for an object.
- ☐ Explain why something modeled as an object can have only kinetic energy, and why a system can have other types of energy.
- ☐ Describe how the work-energy theorem relates to conservation of energy for an object or a system.
- ☐ Explain the meaning of potential energy and how conservative interactions, such as those described by the gravitational and spring forces, give rise to potential energy.
- ☐ Recognize why the work-energy theorem applies even for curved paths and varying forces such as the spring force.

7-1 The ideas of work and energy are intimately related, and this relationship is based on a conservation principle

While you Read the Section: Important Terms and Equations

Use the space below to define each term in your own words. You may also add any other notes that will be helpful for future review.

conservation

conservation law

energy

closed, isolated system

law of conservation of energy

work

potential energy

internal energy

After you Read the Section: Check Your Understanding

Choose the best answer to each of the following. Use the space provided to write a short justification for your selection. When you're finished, check that you got the right answers for the right reasons!

1. Energy is a conserved quantity. This means that

   a. the energy of a system must always be constant.

   b. energy is an important quantity that we should not waste.

   c. the energy of a system can change, but only if energy enters or leaves the system.

   d. the energy of a system can change, but only if energy is created or destroyed in the system.

   Justify your choice:

2. Which of the following statements about energy is true?

   a. A system can have only one kind of energy (for example, kinetic energy) at a time.

   b. Internal energy is the total of all forms of energy in a system.

   c. When energy is converted from one type to another type in a system, some energy is always lost in the process.

   d. The conservation of energy limits what is possible for an object or system.

   Justify your choice:

7-2 The work done by a constant force exerted on a moving object depends on the magnitude of the force and the distance the object moves in the direction of the force

Notice that there are **two equations for work** introduced in this section. The first equation is just a "simpler" form of the second equation, for when the force is in the same direction as the displacement. In such a case, the angle would be zero degrees. Since cos (0°) = 1, the cosine of the angle can be ignored.

Also note that because work is the product of force and displacement, it can be calculated based on the **AREA under a force-position graph**, provided that the force being graphed is parallel to the displacement.

While you Read the Section: Important Terms and Equations

Use the space below to define each term in your own words. For equations, use the space to identify what each letter represents and its associated SI unit. You may also add any other notes that will be helpful for future review.

Joule

$W = Fd$

$W = Fd \cos \theta$

After you Read the Section: Check Your Understanding

Choose the best answer to each of the following. Use the space provided to write a short justification for your selection. When you're finished, check that you got the right answers for the right reasons!

1. Which of the following statements about work is true?

    a. Work can be positive or negative, depending on the angle between the direction of the force exerted on the object and the direction of the object's displacement.

    b. Work can be positive or negative, depending on the direction of the work.

    c. Work is a vector, and therefore can never be negative.

    d. Work is a scalar, and therefore can never be negative.

    Justify your choice:

2. In which one of the following is the total work done by the gravitational force on the object equal to zero?

    a. A person lifts an object upwards from the floor.

    b. A person lifts an object upwards from the floor, and then lowers it back down to the floor.

    c. A person lowers an object down to the floor.

    d. A person drops an object so that it falls down to the floor.

    Justify your choice:

3. How much work is done by the gravitational force on an object that weighs 3 N if it is dragged 4 m across a level floor?

    a. 0 J   b. 1 J   c. 7 J   d. 12 J

    Justify your choice:

7-3 Newton's second law applied to an object allows us to determine a formula for kinetic energy and state the work-energy theorem for an object

The work-energy theorem is extremely important... but the work-energy theorem shown in this section can be used only for idealized **OBJECTS**, which is all that is being considered so far. The work-energy theorem for **SYSTEMS** will be introduced in Section 7-5. If you're not sure about this, now would be a great time to **review objects and systems from pages 8 & 9 of your textbook!**

While you Read the Section: Important Terms and Equations

Use the space below to define each term in your own words. For equations, use the space to identify what each letter represents and its associated SI unit. You may also add any other notes that will be helpful for future review.

work-energy theorem

translational kinetic energy

$$v_f^2 = v_i^2 + 2a_x(x_f - x_i)$$

$$W_{net} = ma_x(x_f - x_i)$$

$$K = \frac{1}{2}mv^2$$

$$W_{net} = K_f - K_i$$

After you Read the Section: Check Your Understanding

Choose the best answer to each of the following. Use the space provided to write a short justification for your selection. When you're finished, check that you got the right answers for the right reasons!

1. Two objects are moving at the same speed v_0. The first object has mass m_1 and the second object has mass $m_2 = 2m_1$. Which of the following correctly relates their kinetic energies?

   a. $K_1 = 0.5\,K_2$ b. $K_1 = 2\,K_2$ c. $K_1 = 2\,K_2$ d. $K_1 = 4\,K_2$

   Justify your choice:

2. Two objects have the same mass. The first object moves with speed v_1 and the second object moves with speed $v_2 = 0.5\,v_1$. Which of the following correctly relates their kinetic energies?

   a. $K_1 = 0.5\,K_2$ b. $K_1 = 2\,K_2$ c. $K_1 = 2\,K_2$ d. $K_1 = 4\,K_2$

   Justify your choice:

3. How fast does a 4 kg object need to be moving to have 50 J of kinetic energy?

   a. 2 m/s   b. 5 m/s   c. 12.5 m/s   d. 25 m/s

   Justify your choice:

7-4 The work-energy theorem can simplify many physics problems

This section extends the ideas introduced in earlier sections by applying them to a wider range of situations, so there are no new terms or equations introduced here. **It is important to reach a good understanding of these first four sections before continuing with the rest of the chapter.**

After you Read the Section: Check Your Understanding

Choose the best answer to each of the following. Use the space provided to write a short justification for your selection. When you're finished, check that you got the right answers for the right reasons!

After reading this section and trying the questions below, I recommend that you then **work on the Chapter 7 Review Problems 1 to 9** before continuing further into the chapter.

1. A 2 kg object initially moving at 2 m/s accelerates to 3 m/s. What was the net work done to the object?

    a. 2 J   b. 4 J   c. 5 J   d. 9 J

    Justify your choice:

2. Objects 1 and 2 have equal mass. Both objects are initially at the same height above the floor, and both objects end up on the floor. Object 1 was dropped and fell straight to the floor, but object 2 was released on an incline and slid down to the floor. How does the work done by the force of gravity to the two objects, W_1 and W_2, compare?

    a. $W_1 = W_2$ b. $W_1 > W_2$ c. $W_1 < W_2$

    d. The comparison cannot be made without knowing if the surface had negligible friction or not.

    Justify your choice:

3. An object is initially moving at 2 m/s to the right. Two forces are then exerted on the object, after which the object is moving at 2 m/s to the left. If the first force did 30 J of work, then how much work was done by the second force?

    a. 0 J   b. 30 J   c. -30 J   d. -60 J

    Justify your choice:

4. A block is dropped from rest directly above a pillow. The block falls freely towards the pillow, hits the pillow, and is brought to a stop a distance d_1 below the point from which it was released. The block moved a distance d_2 while it was in contact with the pillow. Which of the following is a correct expression for the magnitude of the average force exerted on the block by the pillow while the pillow was slowing to a stop?

    a. mg b. mgd_1d_2 c. mgd_2/d_1 d. mgd_1/d_2

    Justify your choice:

7-5 The work-energy theorem is also valid for curved paths and varying forces, and, with a little more information, systems as well as objects

The energy aspect of springs is introduced in this section. It may be helpful to review what has already been introduced about springs—especially **Hooke's Law**—from Section 5-6 before reading this section!

This section also begins to apply the ideas of work and energy (which have so far been applied only to the **object model**) to more complicated **systems**. When answering questions in the future, always be mindful of which model is needed as the object model is simpler, but more limited.

Don't confuse the symbols for kinetic energy (upper-case "K") and spring constant (lower-case "k"). When writing equations with both symbols, go out of your way to make your handwritten k's look distinct from K's, OK?

While you Read the Section: Important Terms and Equations

Use the space below to define each term in your own words. For equations, use the space to identify what each letter represents and its associated SI unit. You may also add any other notes that will be helpful for future review.

$$W = F d_{contact} \cos(\theta)$$

$$W = \Delta E = \Delta K + \Delta U + \Delta E_{internal}$$

$$W = \frac{1}{2} k x_2^2 - \frac{1}{2} k x_1^2$$

After you Read the Section: Check Your Understanding

Choose the best answer to each of the following. Use the space provided to write a short justification for your selection. When you're finished, check that you got the right answers for the right reasons!

After reading this section and trying the questions below, I recommend that you then **work on the Chapter 7 Review Problems 10 to 14** before continuing further into the chapter.

1. Which of the following best explains why the work-energy theorem is sometimes written as $W_{net} = K_f - K_i$, and other times as $W = \Delta E = \Delta K + \Delta U + \Delta E_{internal}$?

    a. The smaller equation can be used only if air resistance and friction are negligible.

    b. The smaller equation applies only to the object model and the larger equation can be used for more complicated systems.

    c. Both of the above are true.

    d. Neither of the above are true.

    Justify your choice:

2. A person is initially crouched on the floor with their knees bent. They then jump upwards by straightening their legs. While doing so, the normal force exerted on the person by the floor is 600 N. The person's head moves upwards 50 cm while still in contact with the floor, then after leaving the floor the person continues to rise an additional 50 cm before reaching their maximum height. How much work is done by the normal force on the person?

    a. 0 J   b. 300 J   c. 600 J   d. 900 J

    Justify your choice:

7-6 Potential energy is energy related to reversible changes in a system's configuration

Students are often initially confused by the distinction between **conservative forces** and **nonconservative forces**. It's okay if you are confused by this for now—your understanding will improve if you keep working at it, especially as these ideas will continue to be used in Chapter 8.

Also note that in this section the idea of the **system model** is used extensively. It is critical that you either define what is in the system being considered, or if a question has defined the system for you, that you answer the question based on that system. **Don't jump into using any equations until you are clear about what is in the system!**

While you Read the Section: Important Terms and Equations

Use the space below to define each term in your own words. For equations, use the space to identify what each letter represents and its associated SI unit. You may also add any other notes that will be helpful for future review.

conservative force

nonconservative force

gravitational potential energy

spring potential energy

$U_{grav} = mgy$

$U_s = \dfrac{1}{2}kx^2$

After you Read the Section: Check Your Understanding

Choose the best answer to each of the following. Use the space provided to write a short justification for your selection. When you're finished, check that you got the right answers for the right reasons!

After reading this section and trying the questions below, I recommend that you then **work on the Chapter 7 Review Problems 15 to 26** before continuing further into the chapter.

1. A person lifts a 1-kg object up a distance of 2 m. How much work is done **by the person to the object**?

    a. 0 J   b. -20 J   c. 20 J

    Justify your choice:

2. A person lifts a 1-kg object up a distance of 2 m. How much work is done **by the person to the Earth-object system**?

    a. 0 J   b. -20 J   c. 20 J

    Justify your choice:

3. A person lifts a 1-kg object up a distance of 2 m. How much work is done **by the person to the Earth-object-person system**?

    a. 0 J   b. -20 J   c. 20 J

    Justify your choice:

4. A person lifts a 1-kg object up a distance of 2 m. How much work is done **by the force of gravity to the object**?

    a. 0 J   b. -20 J   c. 20 J

    Justify your choice:

5. A person lifts a 1-kg object up a distance of 2 m. How much work is done **by the force of gravity to the Earth-object system**?
   a. 0 J
   b. -20 J
   c. 20 J

   Justify your choice:

6. A 1-kg object is 2 meters off the floor. How much potential energy does the **object** have?
   a. 0 J
   b. -20 J
   c. 20 J

   Justify your choice:

7. A 1-kg object is 2 meters off the floor. How much potential energy does the **object-Earth system** have?
   a. 0 J
   b. -20 J
   c. 20 J

   Justify your choice:

8. Considering only the spring force F_s, friction force F_f, and tension force, F_T, which of these forces is/are conservative?
   a. only F_s
   b. only F_f
   c. only F_T
   d. None of them

   Justify your choice:

After you Read the Chapter: Test Yourself

After reading the chapter and trying the questions below, I recommend that you then **work on the Chapter 7 AP Practice Problems** at the end of the chapter in your textbook before continuing into the next chapter.

Word Match

A) Closed, Isolated System B) Conservation C) Conservative Force D) Dissipate

E) Energy F) External Force G) Hooke's Law H) Ideal Spring

I) Internal Energy J) Joule K) Kinetic Energy L) Mechanical Energy

M) Net Work N) Nonconservative O) Object P) Potential Energy

Q) Spring Constant R) Spring Potential Energy S) System T) Work

U) Work-Energy Theorem

___ 1. Energy related to mass being in motion.

___ 2. Can only have kinetic energy and no other form of energy.

___ 3. Each one of these has an associated potential energy.

___ 4. Net work changes the kinetic energy of an object.

___ 5. Is transferred by work.

___ 6. Includes thermal energy.

___ 7. The mechanical transfer of energy by an external force.

___ 8. Includes everything that is not in the environment.

___ 9. Has negligible mass and can store potential energy.

___ 10. Can change the energy of a system by doing work to it.

___ 11. Interactions within a system that change mechanical energy of the system into other forms.

___ 12. Has no work done to it by external forces.

___ 13. Includes kinetic energy and potential energy.

___ 14. Work done by all external forces.

___ 15. When energy becomes spread out and difficult to find or to use.

___ 16. A general principle indicating that something can't be created or destroyed.

___ 17. Energy that can readily be turned into kinetic energy.

___ 18. The unit of both work and energy.

___ 19. Energy stored in a spring that is not relaxed.

___ 20. Measure of the stiffness of a spring.

___ 21. Equation for finding force provided by a spring.

End of Chapter Multiple Choice Questions

1. A moving object has kinetic energy K_0. Its speed doubles, and then that higher speed is tripled. What is its final kinetic energy in terms of K_0?

   a. $5 K_0$ b. $13 K_0$ c. $25 K_0$ d. $36 K_0$

2. The graph here shows the net force exerted on a 3.0 kg object that started from rest. What is the kinetic energy of the object at time $t = 4.0$ s?

   a. 0 J
   b. 8.0 J
   c. 24 J
   d. 48 J

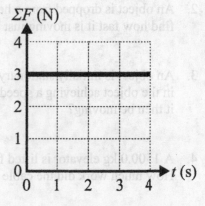

3. A pendulum bob swings from its highest point to its lowest point. Which of the following is correct regarding the work done to the pendulum bob throughout this motion by the force of tension (W_T) and the work done by the force of gravity (W_g)?

   a. $W_g > 0; W_T = 0$ b. $W_g < 0; W_T = 0$ c. $W_g = 0; W_T > 0$ d. $W_g = 0; W_T < 0$

4. Three objects are thrown out of the window of a tall building with the same initial speeds: object #1 is thrown vertically downwards, object #2 is thrown vertically upwards (it rises up before falling to the ground below), and object #3 is thrown horizontally (it moves in a parabolic arc to the ground below). Which of the following correctly ranks the speeds of the three objects just before each hits the ground?

   a. $v_1 > v_2 > v_3$ b. $v_1 = v_2 = v_3$ c. $v_1 = v_2 > v_3$ d. $v_2 > v_1 > v_3$

5. A person holds an inflated balloon between her hands. She then squeezes the balloon by pushing both of her hands closer together. In doing so, her hands each exert an average force of 30 N onto the balloon, and each hand moves 5 cm. The center of the balloon does not move at all. How much work did she do to the balloon?

   a. 0 b. 1.5 J c. 3 J d. 6 J

End of Chapter Problems

Solve each of the following problems on separate paper.

1. Two people simultaneously push a box on the floor. The first person pushes the box west with a force of 250.0 N while the second person pushes it with a force of 75.0 N east. The box slides 5.00 meters to the west. What was the net work done to the box by the people?

2. An object is dropped from a height of 2.00 m above the ground. Use the ideas of work and energy to find how fast it is moving just before it hits the ground.

3. An object is initially stationary. It then has a net work of 50.0 J done to it by external forces, resulting in the object achieving a speed of 15.0 m/s. If an additional 50.0 J of work is then done, how fast will it then be moving?

4. A 1500.0 kg elevator is lifted from rest by a cable. After rising 3.00 meters, it is moving at 2.00 m/s. How much work did the cable do to the elevator in this lift?

5. A person pushes a 15.0 kg crate 8.00 m across the floor by pushing on it with a force of 200.0 N. The crate was initially at rest, but after this push it is moving at 3.00 m/s. How much work was done on the crate by the force of friction?

6. A "human cannon" is made by placing a large spring inside of a tube. The spring is 3.00 m long when relaxed. The tube is oriented vertically with the lower end of the spring at ground level. A 75.0 kg person then stands on top of the spring and is gently lowered down. When the spring is 2.50 m long, it can fully support the weight of the person. The spring is then forced down until it is 1.20 m long at which point it and the person on top of it are released. The spring quickly extends back to its relaxed length at which point the person gets launches upward. What is the maximum height above the ground the person will reach?

Chapter 8
Application of Conservation Principles

Before you Read the Chapter: Prepare Yourself

Be sure that you have a good understanding of:

- **Chapter 7**! Don't start Chapter 8 until you have a good understanding of Chapter 7.

Chapter Overview

Chapter 8 is a direct continuation of the material that was introduced in Chapter 7. The only new concept introduced in this chapter is power. In addition, Chapter 8 reinforces the ideas that were introduced in Chapter 7 and explores them more deeply. It also adds a "general" approach to gravitational potential energy, as it turns out that the approach introduced in Chapter 7 is limited (and so it can't always be used).

Learning Objectives

☐ Identify which kinds of problems are best solved with energy conservation and the steps to follow in solving these problems.

☐ Describe how changing the way you identify a system changes the description of conservation of energy (but not the results).

☐ Identify the types of energy involved in an interaction between objects and systems.

☐ Describe what power is and its relationship to work and energy.

☐ Describe the general expression for gravitational potential energy and how to relate it to the expression used near Earth's surface.

8-1 Total energy is always conserved, but it is constant only for a closed, isolated system

As usual, the first section of this chapter is introductory, and lays the foundation for important ideas that will be expanded on later in the chapter. Conservation of energy was introduced in Chapter 7, but here in Chapter 8 this idea will be refined, so it is worth paying attention to every detail.

While you Read the Section: Important Terms and Equations

Use the space below to define each term in your own words. You may also add any other notes that will be helpful for future review.

mechanical energy

mechanical energy transfer

total mechanical energy

After you Read the Section: Check Your Understanding

Choose the best answer to each of the following. Use the space provided to write a short justification for your selection. When you're finished, check that you got the right answers for the right reasons!

1. Which one of the following best describes the condition that guarantees that the energy of a system is constant?

    a. the system includes only one object

    b. the system has only kinetic energy

    c. the system is closed and isolated

    d. the energy in the system is conserved

    Justify your choice:

2. If a system is closed and isolated, which one of the following must be true about the system?

    a. No energy can enter or leave the system, so the energy of the system must be decreasing.

    b. No energy can enter or leave the system, so the energy of the system must be constant.

    c. No energy can enter or leave the system, so the energy of the system must be increasing.

    d. Energy can enter or leave the system, so the energy of the system must be changing.

    Justify your choice:

3. If a closed, isolated system has constant mechanical energy, which one of the following best describes interactions within the system?

    a. There must be interactions occurring within the system.

    b. There must be no interactions occurring within the system.

    c. If any interactions occur within the system, they must be nonconservative interactions.

    d. If any interactions occur within the system, they must be conservative interactions.

    Justify your choice:

8-2 Choosing systems and considering multiple interactions, including nonconservative ones

This section reinforces the ideas introduced in the first section along with ideas from Chapter 7. Work-Energy bar diagrams are also introduced. These diagrams can be helpful when trying to visualize the energy of a system and how it can change.

While you Read the Section: Important Terms and Equations

Use the space below to define each term in your own words. For equations, use the space to identify what each letter represents and its associated SI unit. You may also add any other notes that will be helpful for future review.

thermal energy

$$K_i + U_i = K_f + U_f$$

$$\Delta K + \Delta U = 0$$

$$-\Delta E_{internal} = W_{nonconservative} = -F_{friction} d$$

After you Read the Section: Check Your Understanding

Choose the best answer to each of the following. Use the space provided to write a short justification for your selection. When you're finished, check that you got the right answers for the right reasons!

1. A system initially contains 5 joules of energy. If 3 joules of energy are added to the system, then 2 joules of energy are removed from the system, and then an internal interaction transforms 1 joule of kinetic energy into internal energy, how much energy does the system then have?

    a. 5 J   b. 6 J   c. 7 J   d. 8 J

    Justify your choice:

2. A system is closed and isolated and has only conservative interactions in it. It initially has 2 joules of kinetic energy and 3 joules of potential energy. At a later time, it has 5 joules of kinetic energy. Which one of the following best describe what happened?

    a. Three joules of potential energy were converted to kinetic energy within the system.
    b. Three joules of energy were gained by the system.
    c. Three joules of internal energy were lost by the system.
    d. Energy must not be conserved in the system.

    Justify your choice:

3. A system is closed and isolated. It initially has 6 joules of kinetic energy and 4 joules of potential energy. At a later time, it has 7 joules of kinetic energy and 5 joules of potential energy. Which one of the following best describe what happened?

    a. The system must have had 2 joules of energy transferred into it.
    b. The system must have had 2 joules of energy transferred out of it.
    c. The system must have lost 2 joules of internal energy
    d. The system must have gained 2 joules of internal energy

    Justify your choice:

8-3 Energy conservation is an important tool for solving a wide variety of problems

This section wraps up the topic of energy conservation by applying the ideas from the previous sections and the material introduced earlier to a wide range of example problems. While looking over the examples it is important that you not merely attempt to "memorize the steps" but rather reach an understanding of how the problem is solved so that you will be able to solve a wide range of different problems.

This section does not introduce any new terms or equations.

After you Read the Section: Check Your Understanding

Choose the best answer to each of the following. Use the space provided to write a short justification for your selection. When you're finished, check that you got the right answers for the right reasons!

1. Which one of the following is an example of an advantage that solving problems with the law of conservation of energy has, compared to solving problems with Newton's laws and kinematics?

    a. More information can be provided with work and energy because they are vectors.

    b. Less information is needed to solve problems with work and energy because they are scalars.

    c. Work and energy are more useful as kinematics can be used only if there is no net force exerted on a system.

    d. None of the above.

    Justify your choice:

2. The choice of which objects are included in a system can affect which of the following?

    a. the amount of energy in the system

    b. whether a force is classified as being internal or external

    c. whether or not the energy in the system is constant

    d. all of the above

    Justify your choice:

3. A force can do work to a system only if it is

    a. External and has a component parallel to the motion

    b. External and has a component perpendicular to the motion

    c. Internal and conservative

    d. Internal and nonconservative

    Justify your choice:

8-4 Power is the rate at which energy is transferred into or out of a system or converted within a system

This section introduces power, which is a useful quantity for measurement. For example, motors are often described in terms of power (though other measurements would also be needed to give a more complete description). Note that power is another word that has a formal definition in physics that is not quite the same as how the word is often used outside of physics, so be sure to reach a good understanding of how the word is being used here.

While you Read the Section: Important Terms and Equations

Use the space below to define each term in your own words. For equations, use the space to identify what each letter represents and its associated SI unit. You may also add any other notes that will be helpful for future review.

Power

Watt

$$P = \frac{\Delta E}{\Delta t}$$

$$P = \frac{W}{\Delta t} = \frac{(F\cos\theta)d}{\Delta t} = (F\cos\theta)v$$

Chapter 8 | Application of Conservation Principles

After you Read the Section: Check Your Understanding

Choose the best answer to each of the following. Use the space provided to write a short justification for your selection. When you're finished, check that you got the right answers for the right reasons!

1. A motor does 50 joules of work in 2 seconds. What was the power output of the motor?

    a. 25 J   b. 50 J   c. 25 W   d. 100 W

    Justify your choice:

2. A battery supplies 15 W for 2 minutes. How much energy does the battery provide in doing so?

    a. 30 J   b. 900 J   c. 1800 J   d. 3600 J

    Justify your choice:

3. A person carries a 2.0 kg object 3.0 meters at constant speed for 5.0 s. The person exerts an upward force on the object to support it while moving the object horizontally. What is the power output of the person?

    a. 0   b. 1.2 W   c. 12 W   d. 30 W

    Justify your choice:

4. A block slides on a horizontal surface to the east at a constant speed of 4 m/s. A 5 N force that is directed at 60° north of east is exerted on the object. What is the power delivered to the object by this force?

    a. 0   b. 10 W   c. 20 W   d. 40 W

    Justify your choice:

8-5 Gravitational potential energy is much more general, and profound, than our near-Earth approximation

This section reveals that the equation introduced in Chapter 7 for gravitational potential energy is limited to situations in which the acceleration due to gravity g is constant (or acceptably approximated to be constant), such as when an object stays close to the surface of Earth. In situations in which an object does not remain close to the surface of Earth, the equation introduced in this section is needed.

While you Read the Section: Important Terms and Equations

Use the space below to define each term in your own words. For equations, use the space to identify what each letter represents and its associated SI unit. You may also add any other notes that will be helpful for future review.

escape speed

$$U_{grav} = -\frac{Gm_1 m_2}{r}$$

$$v_{escape} = \sqrt{\frac{2Gm_{Earth}}{R_{Earth}}}$$

After you Read the Section: Check Your Understanding

Choose the best answer to each of the following. Use the space provided to write a short justification for your selection. When you're finished, check that you got the right answers for the right reasons!

1. Which of the following best describes the condition needed for the gravitational potential energy of an object-Earth system to be reasonably calculated with the equation $U_{grav} = mgy$?

    a. The object must have a mass m that is much less than the mass of Earth.

    b. The object must initially be at $y = 0$ on the ground.

    c. The object must remain close to the surface of Earth.

    d. The object must be stationary.

    Justify your choice:

2. When analyzing situations in which an object does not remain close to the surface of Earth, which of the following best describes the condition needed for the object-Earth system to have zero gravitational potential energy?

    a. The object must be stationary.

    b. The object must be on the ground.

    c. The object must be close to the surface of Earth.

    d. The object must be infinitely far from Earth.

    Justify your choice:

3. The system composed of Earth and an object that was far from Earth's surface had -100 J of potential energy. The object remained stationary, but Earth moved so that the distance between them doubled. How much potential energy does the system now have?

    a. -50 J   b. -100 J   c. -200 J   d. 0

    Justify your choice:

4. A moon has radius R_0. An object launched from the surface of the moon has escape speed v_1. If the object was instead launched from a position that is a height R_0 above the surface of the moon, the object has escape speed v_2. Which of the following are true about v_1 and v_2?

    a. $v_2 = v_1$ b. $v_2 = v_1 / 2$ c. $v_2 = \sqrt{2} v_1$ d. $v_2 = v_1 / \sqrt{2}$

    Justify your choice:

After you Read the Chapter: Test Yourself

After reading the chapter and trying the questions below, I recommend that you then **work on the Chapter 8 Review Problems** at the end of the chapter in your textbook before continuing into the next chapter.

End of Chapter Multiple Choice Questions

1. A block is placed on top of a vertical spring. The block is initially at rest and the spring is relaxed. A person then slowly lowers the block down until its weight is supported at rest by only the spring. For which of the following systems was the mechanical energy of the system constant?

   a. block-spring
   b. block-Earth
   c. block-spring-Earth
   d. none of these systems

2. Two objects with masses m_1 and m_2 ($m_2 > m_1$) are connected by a string which is placed over a pulley as shown here. Both masses are released from rest. Which of the following is a correct expression for the kinetic energy of the system composed of both masses when m_2 has moved down a distance d? Assume that the string and pulley have negligible mass, the string does not stretch and the pulley spins freely with negligible friction.

   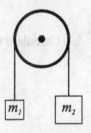

   a. $\frac{1}{2}(m_1 + m_2)gd$
   b. $(m_1 + m_2)gd$
   c. $(m_2 - m_1)gd$
   d. $(m_1 - m_2)gd$

3. An object is dropped near the surface of Earth at $t = 0$. Which of the following graphs best represents the expected graph of the power provided by the force of gravity to the system consisting of the object?

   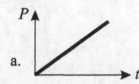 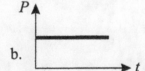  
   a.   b.   c.   d.

4. Three objects, each with mass m, are placed in a row on the x-axis. The first object m_1 is placed at $x_1 = -r$, the second object m_2 is placed at $x_2 = 0$, and the third object m_3 is placed at $x_3 = +r$. The middle object remains at $x_2 = 0$, but external forces are exerted on the other two objects, so that m_1 is moved to a position much farther away on the $-x$ axis, and m_3 is moved to a position much farther away on the $+x$ axis. Which of the following is an expression for the net work done by these external forces to the system of the three objects?

   a. $2\dfrac{Gm^2}{r}$
   b. $\dfrac{5}{2}\dfrac{Gm^2}{r}$
   c. $3\dfrac{Gm^2}{r}$

   d. No net work was done to this system.

5. The escape speed for an object with mass m_0 from the surface of a planet is v_0. What is the escape speed for an object with mass $2m_0$ the surface of the same planet?

   a. $\dfrac{1}{\sqrt{2}}v_0$
   b. v_0
   c. $\sqrt{2}v_0$
   d. $2v_0$

End of Chapter Problems

Solve each of the following problems on separate paper.

1. A motor does work at the rate of 150.0 W. How long would it take this motor to lift a 20.0 kg object 4.00 m?

2. A person throws a ball upward from ground level with an initial speed of 11.0 m/s. Find its maximum height.

3. Mild-mannered physics teacher by day, witty misconception fighter by night (as well as by day), the mighty superhero Mr. Physics throws a ball upwards from ground level with an initial speed of 11,000 m/s. Find its maximum height.

4. Calculate the escape speed from the surface of Earth.

Chapter 9
Momentum, Collisions, and Center of Mass

Before you Read the Chapter: Prepare Yourself

Be sure that you have a good understanding of:
- Center of mass from Section 4-4
- Energy and conservation of energy from Chapters 7 and 8

Chapter Overview

Like energy, momentum is a conserved quantity, so some of the ideas in Chapter 9 are analogous to ideas introduced in Chapters 7 and 8. Unlike energy, however, momentum is a vector quantity, so we will once again need to be aware of directions and work with components of motion.

Learning Objectives

☐ Use the concepts of momentum, center of mass, and system to predict the behavior of objects in everyday situations.

☐ Define the linear momentum of an object and explain how it differs from kinetic energy.

☐ Explain the conditions under which the total momentum of a system is constant and why total momentum is constant in a collision.

☐ Identify the differences and similarities between elastic, inelastic, and completely inelastic collisions.

☐ Apply conservation of momentum and mechanical energy to problems involving elastic collisions.

☐ Relate the momentum change of an object, the force that causes the change, and the time over which the force is exerted on the object.

☐ Find the center of mass of a system and describe how the net force on a system affects the motion of the system's center of mass.

9-1 Newton's third law will help lead us to the idea of momentum

Newton's third law describes one aspect of how objects and systems interact with each other and is a direct result of the conservation of momenta. This section sets the stage by reminding you of this and introducing momentum.

While you Read the Section: Important Terms and Equations

Use the space below to define each term in your own words. You may also add any other notes that will be helpful for future review.

momentum

law of conservation of momentum

After you Read the Section: Check Your Understanding

Choose the best answer to each of the following. Use the space provided to write a short justification for your selection. When you're finished, check that you got the right answers for the right reasons!

1. Which one of the following is best identified as Newton's third law?
   a. The acceleration of an object depends on the net force exerted on it and the object's mass.
   b. When object 1 exerts force on object 2, object 2 exerts an equal magnitude, oppositely directed force on object 1.
   c. Two objects gravitationally attract each other due to the mass that each object has.
   d. If an object has a constant velocity, there must be no net force exerted on the object.

   Justify your choice:

2. The interaction of two objects or systems can be described in terms of
   a. force b. energy c. momentum d. all of these

   Justify your choice:

9-2 Momentum is a vector that depends on an object's mass, speed, and direction of motion

This section introduces the equation for momentum. Whereas kinetic energy depends on mass and the magnitude of velocity (i.e. speed), momentum depends on mass and velocity (including the direction of the velocity). Both the **Watch Out!** box and the **AP® Exam Tip** box on page 384 clarify the important difference between momentum and kinetic energy—be sure to be clear about this difference!

While you Read the Section: Important Terms and Equations

Use the space below to define each term in your own words. For equations, use the space to identify what each letter represents and its associated SI unit. You may also add any other notes that will be helpful for future review.

linear momentum

Collision

$$\vec{a} = \frac{\Delta \vec{v}}{\Delta t} = \frac{\vec{v}_2 - \vec{v}_1}{t_2 - t_1}$$

$$\vec{p} = m\vec{v}$$

After you Read the Section: Check Your Understanding

Choose the best answer to each of the following. Use the space provided to write a short justification for your selection. When you're finished, check that you got the right answers for the right reasons!

1. One object has a mass of 1 kg and moves to the west at 1 m/s while another object also has a mass of 1 kg but moves east at 1 m/s. How do the kinetic energy and momentum of the two objects compare?

    a. The two objects have equal kinetic energy and equal momentum.

    b. The two objects have different kinetic energy and different momentum.

    c. The two objects have equal kinetic energy and different momentum.

    d. The two objects have different kinetic energy and equal momentum.

    Justify your choice:

2. A 2 kg object moves at 4 m/s. How much inertia does the object have?

    a. 2 kg b. 0.5 kg·m/s c. 8 kg·m/s d. 16 J

    Justify your choice:

3. A 2 kg object moves at 4 m/s. How much kinetic energy does the object have?

    a. 2 kg b. 0.5 kg·m/s c. 8 kg·m/s d. 16 J

    Justify your choice:

4. A 2 kg object moves at 4 m/s. What is the magnitude of the momentum of the object?

    a. 2 kg b. 0.5 kg·m/s c. 8 kg·m/s d. 16 J

    Justify your choice:

9-3 The total momentum of a system is always conserved; it is constant for systems that are closed and isolated

This section reveals the main reason why momentum is a useful quantity—it is conserved. Conservation of momentum can be used to solve a wide range of problems and can be combined with conservation of energy to solve even more problems.

While you Read the Section: Important Terms and Equations

Use the space below to define each term in your own words. For equations, use the space to identify what each letter represents and its associated SI unit. You may also add any other notes that will be helpful for future review.

total momentum

$$\left(\Sigma \vec{F}_{\text{external on system}}\right)\Delta t = \vec{p}_{\text{total,f}} - \vec{p}_{\text{total,i}} = \Delta \vec{p}_{\text{total}}$$

$$\vec{p}_{\text{total,f}} = \vec{p}_{\text{total,i}}$$

After you Read the Section: Check Your Understanding

Choose the best answer to each of the following. Use the space provided to write a short justification for your selection. When you're finished, check that you got the right answers for the right reasons!

1. Which one of the following statements is correct for a system that has no external forces exerted on it?

    a. Each part of the system must have constant momentum.

    b. Individual parts of the system can have changing momenta, but the total momentum of the system must be constant.

    c. The total momentum of the system can change, but only if there is one or more internal forces.

    d. The total momentum of the system can change, but only if the system loses internal energy.

    Justify your choice:

2. For a collision, the momentum of the system composed of all colliding objects is usually well-approximated to be the same before and after the collision. Which of the following best explains why this is true?

    a. External forces are usually safe to ignore since the internal forces are typically much larger than any external forces exerted on the system during a collision.

    b. If an object outside of the system exerts an external force onto the system, the system will exert and equal and opposite force on the object.

    c. Internal forces cannot do work to the system.

    d. Any loss of momentum in one part of the system will be balanced by a gain of kinetic energy in another part of the system.

    Justify your choice:

9-4 In an inelastic collision some of the mechanical energy is dissipated

This section introduces four different types of collision with emphasis on inelastic collisions (elastic collisions will be featured in the next section). Remember that while energy is always conserved, mechanical energy can be transformed into other kinds of energy, so mechanical energy itself need not be conserved.

While you Read the Section: Important Terms and Equations

Use the space below to define each term in your own words. For equations, use the space to identify what each letter represents and its associated SI unit. You may also add any other notes that will be helpful for future review.

elastic collision

inelastic collision

explosive collision

completely inelastic collision

$$m_A \vec{v}_{Ai} + m_B \vec{v}_{Bi} = (m_A + m_B)\vec{v}_f$$

After you Read the Section: Check Your Understanding

Choose the best answer to each of the following. Use the space provided to write a short justification for your selection. When you're finished, check that you got the right answers for the right reasons!

1. For which of the following types of collisions is the total momentum of the colliding objects the same before and after the collision?

    a. elastic collisions
    b. inelastic collisions
    c. completely inelastic collisions
    d. all of these

    Justify your choice:

2. For which of the following types of collisions is the total kinetic energy of the colliding objects the same before and after the collision?

    a. elastic collisions
    b. inelastic collisions
    c. completely inelastic collisions
    d. all of these

    Justify your choice:

3. Which of the following correctly describes a completely inelastic collision?

    a. The total kinetic energy of the colliding objects after the collision is greater than their total kinetic energy before the collision.

    b. The total kinetic energy of the colliding objects after the collision is equal to their total kinetic energy before the collision.

    c. The total kinetic energy of the colliding objects after the collision can be nonzero and is always less than their total kinetic energy before the collision.

    d. The total kinetic energy of the colliding objects after the collision is always zero.

    Justify your choice:

4. A 5.00 kg object moving east at 20.0 m/s collides with a 15.0 kg object moving west at 4.00 m/s. What will their speed be after they collide if they then stick together?

    a. 0
    b. 2 m/s
    c. 8 m/s
    d. 12 m/s

    Justify your choice:

132   Section 9-4

9-5 In an elastic collision both momentum and mechanical energy are constant

This section does not introduce any new terms or equations. Instead, it focuses on elastic collisions. This type of collision is special as the system of colliding objects will not just have the same momentum before and after the collision, but will also have the same kinetic energy before and after the collision. As both of these quantities are constant, it opens up the possibility of answering problems that require the use of two separate equations (one based on momentum, and one based on kinetic energy). Although this is an important kind of problem that you should certainly understand and be able to solve, the **AP® Exam Tip** box on page 413 correctly points out that you will not be asked to solve complicated simultaneous equations on the AP® exam.

After you Read the Section: Check Your Understanding

Choose the best answer to each of the following. Use the space provided to write a short justification for your selection. When you're finished, check that you got the right answers for the right reasons!

1. Which of the following best describes the total momentum of a system of objects that collide elastically with each other?

    a. The total momentum is constant before, during, and after the collision.

    b. The total momentum is not constant during the collision, but the total momentum after the collision is equal to the total momentum before the collision.

    c. The total momentum is not constant during the collision, and the total momentum after the collision is not equal to the total momentum before the collision.

    d. The total momentum is constant during the collision, but the total momentum after the collision is not equal to the total momentum before the collision.

    Justify your choice:

2. Which of the following best describes the total kinetic energy of a system of objects that collide elastically with each other?

    a. The total kinetic energy is constant before, during, and after the collision.

    b. The total kinetic energy is not constant during the collision, but the total kinetic energy after the collision is equal to the total kinetic energy before the collision.

    c. The total kinetic energy is not constant during the collision, and the total kinetic energy after the collision is not equal to the total kinetic energy before the collision.

    d. The total kinetic energy is constant during the collision, but the total kinetic energy after the collision is not equal to the total kinetic energy before the collision.

    Justify your choice:

3. A 2 kg object moving east at 4 m/s collides with a 4 kg object that was initially stationary. After the collision the 2 kg object is stationary and the 4 kg object is moving east at 2 m/s. Based on this information, which of the following correctly identifies and explains the type of collision that occurred?

   a. Elastic, because the objects bounced off of each other without sticking to each other.
   b. Elastic, because both total momentum and total kinetic energy were the same before and after the collision.
   c. Inelastic, because the total kinetic energy was the same before and after the collision, but the total momentum before and after the collision were not the same.
   d. Inelastic, because the total momentum was the same before and after the collision, but the total kinetic energy before and after the collision were not the same.

   Justify your choice:

4. A 2 kg object moving east at 4 m/s collides with a 6 kg object that was initially stationary. After the collision the 2 kg object is moving west at 2 m/s and the 6 kg object is moving east at 2 m/s. Based on this information, which of the following correctly identifies and explains the type of collision that occurred?

   a. Elastic, because the objects bounced off of each other without sticking to each other.
   b. Elastic, because both momentum and mechanical energy were conserved.
   c. Inelastic, because momentum was not conserved but mechanical energy was conserved.
   d. Inelastic, because momentum was conserved but mechanical energy was not conserved.

   Justify your choice:

9-6 What happens in a collision is related to the time the colliding objects are in contact

This section introduces the concept of impulse, which provides a more complete understanding of the conservation of momentum: in the same way that external forces can do work to change the total energy of a system, we now see that external forces provide an impulse to change the total momentum of a system. Remember that while these are both true, they are not to be confused with each other. It is possible, for example, for an external force to provide an impulse while doing no work! The **AP® Exam Tip** box on page 418 should help you to understand this.

While you Read the Section: Important Terms and Equations

Use the space below to define each term in your own words. For equations, use the space to identify what each letter represents and its associated SI unit. You may also add any other notes that will be helpful for future review.

Impulse

impulse-momentum theorem

contact time

$$\vec{J} = \left(\Sigma \vec{F}_{\text{external on object}}\right)\Delta t = \vec{p}_f - \vec{p}_i = \Delta \vec{p}$$

$$\vec{F}_{\text{collision}} \Delta t = \vec{p}_f - \vec{p}_i = \Delta \vec{p}$$

Chapter 9 | Momentum, Collisions, and Center of Mass 135

After you Read the Section: Check Your Understanding

Choose the best answer to each of the following. Use the space provided to write a short justification for your selection. When you're finished, check that you got the right answers for the right reasons!

1. Which of the following units have the same dimensions?

    a. N/s and kg·m/s
    b. N·s and kg·m/s
    c. N/s and kg·m/s^2
    d. N·s and kg·m/s^2

    Justify your choice:

2. Object 1 has a net external force of 12 N exerted on it for 1 second. Object 2 has a net external force of 3 N exerted on it for 4 seconds. Object 3 has a net external force of 1 N exerted on it for 12 seconds. Which of the following correctly ranks the magnitude of the impulse J supplied to each object?

    a. $J_1 = J_2 = J_3$
    b. $J_1 > J_2 > J_3$
    c. $J_1 < J_2 < J_3$
    d. $J_1 > J_2 < J_3$

    Justify your choice:

3. A 3 kg object moves east at 4 m/s. Which of the following would bring the object to a stop?

    a. A net force of 12 N to the west exerted for 1 second
    b. A net force of 3 N to the west exerted for 4 seconds
    c. A net force of 1 N to the west exerted for 12 seconds
    d. All of the above

    Justify your choice:

4. An egg dropped to the floor breaks. Another identical egg dropped from the same height lands on a cushion without breaking. Which of the following best explains one factor for why the first egg broke but the second egg did not break?

    a. The egg that hit the floor was stopped by a larger impulse than the egg that hit the cushion.
    b. The egg that hit the floor was stopped by a smaller impulse than the egg that hit the cushion.
    c. The impulse provided to both eggs to stop them was the same, but the cushion exerted a smaller force for a longer time.
    d. The impulse provided to both eggs to stop them was the same, but the cushion exerted a larger force for a shorter time.

    Justify your choice:

9-7 The center of mass of a system moves as though all the system's mass were concentrated there

Center of mass was first introduced in Chapter 2, and then expanded upon in Chapter 4. Both earlier introductions were, however, qualitative (i.e. descriptive only). This section introduces a quantitative (i.e. numeric) way to describe and calculate the position of the center of mass for a system consisting of two or more individual objects.

While you Read the Section: Important Terms and Equations

Use the space below to define each term in your own words. For equations, use the space to identify what each letter represents and its associated SI unit. You may also add any other notes that will be helpful for future review.

weighted average

$$x_{CM} = \frac{m_1}{M_{tot}}x_1 + \frac{m_2}{M_{tot}}x_2 + \frac{m_3}{M_{tot}}x_3 + \ldots + \frac{m_N}{M_{tot}}x_N = \frac{1}{M_{tot}}\sum_{i=1}^{N} m_i x_i$$

$$\vec{p}_{total} = \sum_{i=1}^{N} m_i \vec{v}_i = M_{tot} \vec{v}_{CM}$$

$$\Sigma \vec{F}_{external\ on\ system} = M_{tot} \frac{\Delta \vec{v}_{CM}}{\Delta t} = M_{tot} \vec{a}_{CM}$$

After you Read the Section: Check Your Understanding

Choose the best answer to each of the following. Use the space provided to write a short justification for your selection. When you're finished, check that you got the right answers for the right reasons!

1. A 2.0 kg object is at $x = 5.0$ m and a 3.0 kg object is at $x = 10.0$ m. What is the position of the center of mass of the system of both objects?

    a. 7.0 m b. 7.5 m c. 8.0 m d. 8.5 m

    Justify your choice:

2. A 2 kg object moves east at 10 m/s away from an 8 kg object that is at rest. What is the magnitude of the velocity of the center of mass of the system of both objects?

    a. 2 m/s b. 4 m/s c. 5 m/s d. 10 m/s

    Justify your choice:

3. A net external force of 12 N to the east is exerted on a 6 kg object and a net external force of 28 N to the west is exerted on a 14 kg object. What is the acceleration of the center of mass of the system of both objects?

    a. 0 b. 0.2 m/s² [west] c. 0.2 m/s² [east] d. 0.8 m/s² [west]

    Justify your choice:

After you Read the Chapter: Test Yourself

After reading the chapter and trying the questions below, I recommend that you then **work on the Chapter 9 Review Problems** at the end of the chapter in your textbook before continuing into the next chapter.

End of Chapter Multiple Choice Questions

1. Two objects that have masses $m_1 > m_2$ move with equal magnitudes of momentum $p_1 = p_2$. How do their kinetic energies compare?

    a. $K_1 < K_2$
    b. $K_1 = K_2$
    c. $K_1 > K_2$
    d. More information is needed.

2. An object initially at rest is dropped. It falls and so it gains momentum. Which of the following best explains how momentum is conserved throughout this motion?

    a. The system composed of the object and Earth loses potential energy as the object falls.

    b. In the system of the object and Earth, Earth accelerates upward with equal and opposite momentum. The system has constant momentum.

    c. The falling object eventually hits the ground where it again comes to rest. Momentum is conserved as the final momentum and initial momentum are both zero.

    d. None of the above, as momentum is not conserved in this case.

3. A 2.00 kg object is initially moving at 3.00 m/s. The object then has a net force in the same direction in which it is moving for 8.00 seconds. The magnitude of the net force for the interval is shown on the graph here. What is the object's velocity at the end of this interval?

    a. 11.0 m/s
    b. 15.0 m/s
    c. 19.0 m/s
    d. 25.0 m/s

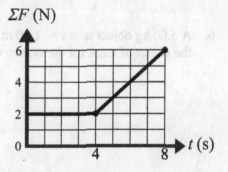

4. An object with mass m_0 was initially moving east with speed v_0. The object then has a completely inelastic collision with a second object with mass $2m_0$ which was initially moving west with speed v_0. What is the speed of the objects after the collision?

    a. 0
    b. v_0
    c. $v_0/2$
    d. $v_0/3$

5. Benjamin and Samantha are both initially at rest facing each other 5.00 meters apart with initial positions $x_B = 0$ and $x_S = 5.00$ m. Since both are on roller skates, the friction with the ground is negligible. Benjamin is holding a heavy ball. The center of mass for the system of Benjamin, Samantha, and the ball is at $x_{CM,i} = 2.00$ m. Benjamin then throws the ball to Samantha. What is the position of the center of mass $x_{CM,f}$ at the moment that Samantha catches the ball?

    a. $x_{CM,f} < 2.00$ m
    b. $x_{CM,f} = 2.00$ m
    c. $x_{CM,f} > 2.00$ m
    d. More information is needed

End of Chapter Problems

Solve each of the following problems on separate paper.

1. An 8.00 kg object is moving at 5.00 m/s to the east. A net force of 12.0 N [east] is then exerted on the object for 4.00 seconds. What is the resulting velocity of the object?

2. A person holds a loaded paintball marker. The person-marker system has a total mass of 100.0 kg. Inside the marker is a paintball which has a mass of 0.00300 kg. Before the person shoots the paintball, everything is at rest. The person then shoots, resulting in the paintball leaving the marker with a speed of 75.0 m/s. Determine the recoil speed of the person (still holding the marker) immediately after shooting the paintball.

3. A 2.00 kg cart rolls along the floor at 6.00 m/s to the east. It strikes a 2.00 kg block which was stationary. The cart bounces back off of the block at 1.2 m/s to the west. Find the velocity of the block after impact and determine the type of collision.

4. A 2.00 kg cart rolls along the floor at 6.00 m/s. It strikes an 8.00 kg block which was stationary. Find the velocity of the block after impact if the collision is completely inelastic.

5. A 2.0 kg cart rolls along the floor at 6.0 m/s to the east. It strikes an 8.0 kg block which was moving north at 3.0 m/s. The collision is completely inelastic. What is their velocity after the collision?

6. A 5.00 kg object is at $x = -1.00$ m. What mass would an object placed at $x = 4.00$ m need in order for the center of mass for the system of both objects to be at $x = 0$?

Chapter 10
Rotational Motion I: A New Kind of Motion

Before you Read the Chapter: Prepare Yourself

Be sure that you have a good understanding of:

- All earlier chapters! (Well, really, chapters 1-8.)

Chapter Overview

Together, Chapters 10 and 11 cover the topic of rotation, which serves as an extension to nearly every topic covered to this point, especially circular motion. Chapter 10 introduces rotational kinematics (extending ideas from Chapters 2 and 3), torque (extending ideas from Chapters 4 and 5), and rotational kinetic energy (extending ideas from Chapters 7 and 8). Chapter 9 (momentum) will be extended in Chapter 11.

Learning Objectives

☐ Define the rotational motion of a system and describe when and how a system can be simplified to an object model.

☐ Define and explain the relationships among angular displacement, angular velocity, and angular acceleration, and use these quantities to describe the rotation of an object or a system of rigidly connected objects, which is a rigid body.

☐ Explain what is meant by the rotational inertia of a system and how to use it to calculate the rotational kinetic energy of that system.

☐ Describe the techniques for finding the rotational inertia of a rigid body, including use of the parallel-axis theorem.

☐ Explain the similarities between rotational and linear motion, and describe how the two forms of motion correspond for a point on a rotating rigid system.

☐ Solve constant-acceleration angular motion problems when given a description of the motion of an object or a system. The description could be in the form of words, graphs, and/or experimental data.

☐ Define the concept of lever arm and know how to use it to calculate the torque caused by a force exerted on a rigid body.

☐ Define rotational equilibrium and calculate it for a rigid body.

☐ Explain how to find the direction of angular velocity and torque.

10-1 Rotation is an important and ubiquitous kind of motion

This section introduces important terminology that will help separate translational motion, which has been the focus of all previous chapters, from rotational motion, which will be the main topic throughout both Chapters 10 and 11.

While you Read the Section: Important Terms and Equations

Use the space below to define each term in your own words. You may also add any other notes that will be helpful for future review.

Translation

extended object

rigid body

Rotation

After you Read the Section: Check Your Understanding

Choose the best answer to each of the following. Use the space provided to write a short justification for your selection. When you're finished, check that you got the right answers for the right reasons!

1. Which of the following kinds of motion are possible?

    a. translation without rotation

    b. rotation without translation

    c. rotation with translation

    d. all of the above

    Justify your choice:

2. A roll of paper is dropped while holding the loose end of the roll so that the paper unrolls as it falls. In doing so, the roll of paper would be best described as which one of the following?

    a. an idealized object

    b. an extended object

    c. a rigid body

    d. an object with negligible size

    Justify your choice:

10-2 An extended object's rotational kinetic energy is related to its angular velocity and how its mass is distributed

As you will see throughout these next two chapters, many aspects of rotational motion are analogous to translational motion. This important section establishes this idea and introduces several such aspects including the rotational equivalents of translational kinematic quantities (e.g., velocity, displacement, and acceleration), inertial mass, and translational kinetic energy.

While you Read the Section: Important Terms and Equations

Use the space below to define each term in your own words. For equations, use the space to identify what each letter represents and its associated SI unit. You may also add any other notes that will be helpful for future review.

angular displacement

angular velocity

angular speed

rotational inertia

rotational kinetic energy

$$\omega_{\text{average},z} = \frac{\Delta \theta}{\Delta t}$$

$$v = r\omega$$

$$I = \sum_{i=1}^{N} m_i r_i^2$$

$$K_{\text{rotational}} = \frac{1}{2} I \omega^2$$

After you Read the Section: Check Your Understanding

Choose the best answer to each of the following. Use the space provided to write a short justification for your selection. When you're finished, check that you got the right answers for the right reasons!

1. An object spins at 180 rpm (revolutions per minute). What is its angular speed?

    a. 3 rad/s   b. 3π rad/s   c. 6 rad/s   d. 6π rad/s

    Justify your choice:

2. A meter stick is oriented vertically with the lower end on the floor. It then falls over, but the lower end on the floor does not move. Which one of the following statements correctly describes its motion while it is falling over?

    a. All points along the length of the meter stick must have the same speed because the meter stick is a rigid body.

    b. All points along the length of the meter stick must have the same speed because the meter stick moves rotationally but not translationally.

    c. Points that are higher up the meter stick must move with larger speed because they are farther from the fixed axis at the lower end of the stick.

    d. Points that are higher up the meter stick must move with smaller speed because they are farther from the fixed axis at the lower end of the stick.

    Justify your choice:

3. A spinning object has 20 J of rotational kinetic energy. How much work must be done to the object to double its angular speed?

    a. 20 J   b. 40 J   c. 60 J   d. 80 J

    Justify your choice:

4. A ball with mass m_1 is sent rolling up an inclined surface. At the bottom of the incline the ball has translational speed v_1 and angular speed ω_1. The ball rolls up the incline without sliding. After translating a distance d_1 it stops moving (both translationally and rotationally) before rolling back down the incline. A block is sent sliding up another surface that is inclined at the same angle. At the bottom of the incline the block has speed $v_2 = v_1$ and slides on the inclined surface with negligible friction. Which of the following correctly describes the distance d_2 the block will slide up the incline?

    a. $d_2 < d_1$ b. $d_2 = d_1$ c. $d_2 > d_1$

    d. The comparison of distance depends on how the mass of the block compares to m_1.

    Justify your choice:

10-3 An extended object's rotational inertia depends on its mass distribution and the choice of rotation axis

Rotational inertia is the angular "version" of inertial mass, but while inertial mass is simply equal to "the mass" of the object, there is more to rotational inertia! I would like to draw your attention to the three properties of rotational inertia described on page 460. These properties are important, so I have included space below for you to summarize them in your own words, along with the usual new terms and equations.

While you Read the Section: Important Terms and Equations

Use the space below to define each term in your own words. For equations, use the space to identify what each letter represents and its associated SI unit. You may also add any other notes that will be helpful for future review.

1st Property of Rotational Inertia:

2nd Property of Rotational Inertia:

3rd Property of Rotational Inertia:

parallel-axis theorem

$I = I_{CM} + Mh^2$

After you Read the Section: Check Your Understanding

Choose the best answer to each of the following. Use the space provided to write a short justification for your selection. When you're finished, check that you got the right answers for the right reasons!

1. A long, thin rod can be rotated about different axes. The diagram here shows three axes: axis 1 goes through the center of the rod along its length, axis 2 also passes through the center of the rod, but is perpendicular to its length, and axis 3 is perpendicular to its length but passes through one end of the rod. Along which of these three axes would the rod have the greatest rotational inertia?

   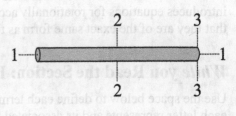

   a. axis 1 b. axis 2 c. axis 3

   d. None of the above; the rotational inertia would be equal for all three axes.

   Justify your choice:

2. Three small objects each with mass m are placed at the three vertices of an equilateral triangle with side length L. What is the rotational inertia of the system of all three objects about an axis that goes through one of the objects, as shown in the diagram here?

   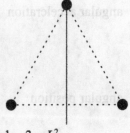

   a. $0.5\, mL^2$ b. $1.5\, mL^2$ c. $2\, mL^2$ d. $3\, mL^2$

   Justify your choice:

3. Three small objects each with mass m are placed at the three vertices of an equilateral triangle with side length L. What is the rotational inertia of the system of all three objects about an axis that goes through one of the objects, as shown in the diagram here?

   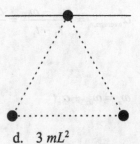

   a. $0.5\, mL^2$ b. $1.5\, mL^2$ c. $2\, mL^2$ d. $3\, mL^2$

   Justify your choice:

10-4 The equations for rotational kinematics are almost identical to those for linear motion

Just as Chapter 2 introduced kinematics equations for translationally accelerated motion, this section introduces equations for rotationally accelerated motion. Looking at these "new" equations should reveal that they are of the exact same form as the more familiar equations we have already been working with.

While you Read the Section: Important Terms and Equations

Use the space below to define each term in your own words. For equations, use the space to identify what each letter represents and its associated SI unit. You may also add any other notes that will be helpful for future review.

rotational kinematics

angular acceleration

angular position

$$\alpha_{\text{average},z} = \frac{\omega_{2z} - \omega_{1z}}{t_2 - t_1} = \frac{\Delta \omega_z}{\Delta t}$$

$$\omega_z = \omega_{0z} + \alpha_z t$$

$$\theta = \theta_0 + \omega_{0z} t + \frac{1}{2} \alpha_z t^2$$

$$\omega_z^2 = \omega_{0z}^2 + 2\alpha_z (\theta - \theta_0)$$

After you Read the Section: Check Your Understanding

Choose the best answer to each of the following. Use the space provided to write a short justification for your selection. When you're finished, check that you got the right answers for the right reasons!

1. An object is initially rotating at 5 revolutions per minute. Its angular speed steadily increases; 3 seconds later it is rotating at 11 revolutions per minute. Although not in SI units, which of the following could be used to correctly state the angular acceleration of the object?

    a. 1 rev / min / s

    b. 2 rev / min / s

    c. 4 rev / min / s

    d. 6 rev / min / s

    Justify your choice:

2. A ceiling fan is rotating at 4π rad/s. It is then shut off. It continues to turn for an additional 30 seconds before coming to a stop. If it slowed with constant angular acceleration, how many revolutions did it complete while coming to a stop?

    a. 2 rev   b. 30 rev   c. 60 rev   d. 120 rev

    Justify your choice:

3. A ball is thrown upwards at 20 m/s, and with an angular speed of 2 rad / s. The ball rises to its highest point, and then returns to be caught at the same height from which it was thrown. What was the ball's angular displacement?

    a. 2 rad   b. 4 rad   c. 6 rad   d. 8 rad

    Justify your choice:

10-5 Torque is to rotation as force is to translation

The title of this section reveals how this section continues to introduce angular quantities that are analogous to translational quantities. Once again, there is more to the angular quantity (torque) than there is to the translational quantity (force).

While you Read the Section: Important Terms and Equations

Use the space below to define each term in your own words. For equations, use the space to identify what each letter represents and its associated SI unit. You may also add any other notes that will be helpful for future review.

torque

line of action

lever arm

rotational equilibrium

right-hand rule

cross product (vector product)

$\tau = rF \sin \phi$

$$\tau = r_\perp F$$

$$\Sigma \tau_{ext,z} = I\alpha_z$$

After you Read the Section: Check Your Understanding

Choose the best answer to each of the following. Use the space provided to write a short justification for your selection. When you're finished, check that you got the right answers for the right reasons!

1. Which of the following statements about force and torque are true?

    a. It is possible to exert a force onto an object without exerting a torque on it.

    b. It is possible to exert a torque onto an object without exerting a force on it.

    c. Both of the above are true.

    d. None of the above are true.

    Justify your choice:

2. A force exerted onto an extended object exerts a torque on the object. The magnitude of the torque could be changed by changing which of the following?

    a. The magnitude of the force.

    b. The direction of the force.

    c. The location on the extended object where the force is applied.

    d. All of the above.

    Justify your choice:

3. An ordinary clock is laying flat on a level table so that a person needs to look downward to read the clock. In which direction is the torque provided to the hands of the clock by the motor, causing them to turn clockwise?

    a. upward   b. downward   c. eastward

    d. It depends on what time it is on the clock.

    Justify your choice:

10-6 The techniques used for solving problems with Newton's second law also apply to rotation problems

Although one new term is introduced, this section primarily provides several examples demonstrating how the ideas introduced earlier in the chapter can be applied to a range of problems.

While you Read the Section: Important Terms and Equations

Use the space below to define the term in your own words. You may also add any other notes that will be helpful for future review.

extended free-body diagram (force diagram)

After you Read the Section: Check Your Understanding

Choose the best answer to each of the following. Use the space provided to write a short justification for your selection. When you're finished, check that you got the right answers for the right reasons!

1. An object with a rotational inertia of 3 kg·m² is free to rotate about a fixed point. If a force F = 8.0 N is exerted on the object as shown with θ = 30°, how far from the axis must the force be exerted to give the object an angular acceleration of 12 rad/s²?

    a. 3 m b. 6 m c. 9 m d. 12 m

    Justify your choice:

2. A ball is released from rest on an inclined surface. The ball accelerates down the incline, rolling without sliding. Which of the following diagrams best represents an extended free-body diagram for the ball? Note that the surface has also been shown as a dashed line.

    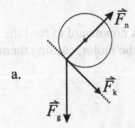 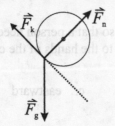 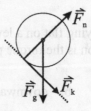 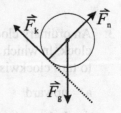

    a. b. c. d.

    Justify your choice:

After you Read the Chapter: Test Yourself

After reading the chapter and trying the questions below, I recommend that you then **work on the Chapter 10 Review Problems** at the end of the chapter in your textbook before continuing into the next chapter.

End of Chapter Multiple Choice Questions

1. An object initially rotating with frequency f_0 slows with constant angular acceleration until it stops rotating, undergoing an angular displacement of θ in the process. Which of the following is an expression for the time it took the object to do this?

    a. $1/f_0$
    b. $\theta/\pi f_0$
    c. $\theta/1\pi f_0$
    d. $\theta/2f_0$

2. A rod is free to rotate about its center and is marked at each quarter of its length. Three forces of equal magnitude F are then exerted on the rod. In which of the following does the rod have the greatest magnitude of net torque from the three forces?

    a.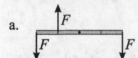
    b.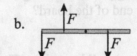
    c.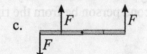
    d.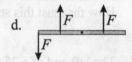

3. The diagram here is a view from above showing a uniform rod that is initially at rest on a level surface. The center of mass of the rod is at its geometric center, and there is negligible friction between the rod and the surface. A horizontal force to the north is then exerted onto the rod at its right end.

    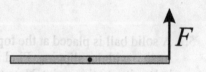

    Which of the following best describes the motion of the rod while the horizontal force is being exerted?

    a. The rod's center of mass will move with constant acceleration to the north, and the rod will initially accelerate rotationally counterclockwise about its center of mass.

    b. The rod's center of mass will remain stationary, and the rod will initially accelerate rotationally counterclockwise about its center of mass.

    c. The rod's right end will move with constant acceleration to the north, and the rod will initially angularly accelerate counterclockwise about its right end.

    d. The rod's center of mass will move with constant acceleration to the north, and the rod will not rotate.

4. A rod with negligible mass has two small objects of equal mass attached to it, one at each end as shown here. The system composed of the rod and two objects has rotational inertia I_{mid} if it is rotated about an axis that passes through the midpoint of the system and has rotational inertia I_{end} if it is rotated about one of the objects at one of the ends of the rod. Which of the following correctly relates I_{mid} to I_{end}?

    a. $I_{mid} = 4I_{end}$
    b. $I_{mid} = 2I_{end}$
    c. $I_{mid} = I_{end}$
    d. $I_{mid} = 0.5I_{end}$

5. Planet 1 rotates about its axis with period T_1 and has rotational kinetic energy K_1. Planet 2 has the same mass and radius as planet 1 but rotates with a period $T_2 = 2T_1$. What is the rotational kinetic energy of the second planet?

   a. $K_1/4$ b. $K_1/2$ c. $2K_1$ d. $4K_1$

End of Chapter Problems

Solve each of the following problems on separate paper.

1. An object is initially rotating clockwise at 8.00 rad/s. It maintains a constant angular velocity for 4.00 seconds and then slows down at a rate of 2.00 rad/s² until it stops rotating. What is the object's angular displacement?

2. A ball has a radius of 0.150 m. It rolls without sliding on the floor at 3.20 m/s. How fast is it rotating?

3. A 75.0 kg person stands on the left end of a 10.0 m long uniform board which is level and pivoted at its center. The 75.0 kg person is balanced by a 125 kg person standing on the right side of the board. How far must this second person be from the right end of the board?

4. The left end of a 25.0 kg uniform pipe is attached to a wall by a pivot. The right end of the pipe is supported by a cable which makes an angle of 30.0° to the pipe. A 50.0 kg object hangs one quarter of the way from its right end. Find the tension in the cable.

5. A solid ball is placed at the top of an inclined plane and released from rest. The ball rolls down the incline without slipping. Find its translational speed at the bottom of the incline, which is 2.00 m below its starting point. Note that a solid ball with mass m and radius r rotating about its center has rotational inertia $I = 0.4mr^2$

Chapter 11
Torque and Rotation II

Before you Read the Chapter: Prepare Yourself

Be sure that you have a good understanding of:

- All ideas from **Chapter 10**
- Satellite Motion from **Chapter 6**
- Gravitational Potential Energy from **Chapter 8**

Chapter Overview

Chapter 10 introduced several aspects of rotational motion including rotational kinematics, dynamics, and kinetic energy. By comparison, Chapter 11 is a shorter chapter that continues where Chapter 10 left off by introducing angular momentum. Kepler's three laws of planetary motion are also included, with the second of these laws in particular serving as an example of angular momentum.

Learning Objectives

☐ Apply the conservation of mechanical energy to rotating systems, including in the case of rolling without slipping.

☐ Describe what is meant by the angular momentum of a rotating system and of a moving object, and explain the circumstances under which angular momentum is constant.

☐ Predict the behavior of rotational collision situations by the same processes that are used to analyze linear collision situations.

☐ Apply the law of universal gravitation and the expression for gravitational potential energy, with the concept of conservation of angular momentum, to analyze the orbits of satellites and planets.

11-1 Angular momentum and the next conservation law: conservation of angular momentum

This section introduces angular momentum, which is the angular version of linear momentum. This is just a warmup, though, as more details including the equation for angular momentum will not be introduced until Section 11-3.

While you Read the Section: Important Terms and Equations

Use the space below to define the term in your own words. You may also add any other notes that will be helpful for future review.

angular momentum

After you Read the Section: Check Your Understanding

Choose the best answer to each of the following. Use the space provided to write a short justification for your selection. When you're finished, check that you got the right answers for the right reasons!

1. Which of the following is the SI unit for linear momentum?

   a. kg·m/s   b. kg·m/s^2 c. kg^2·m/s   d. kg·m^2/s

   Justify your choice:

2. Which of the following is the SI unit for angular momentum?

   a. kg·m/s   b. kg·m/s^2 c. kg^2·m/s   d. kg·m^2/s

   Justify your choice:

3. Which of the following statements about linear momentum and angular momentum is correct?

   a. Linear momentum is conserved, but angular momentum is not conserved.

   b. Linear momentum is not conserved, but angular momentum is conserved.

   c. Linear momentum and angular momentum are each conserved, but separately from each other.

   d. Individually, linear momentum and angular momentum are not conserved, but the total momentum (sum of linear momentum and angular momentum) is conserved.

   Justify your choice:

11-2 Conservation of mechanical energy also applies to rotating extended objects

As we have already seen, both energy and momentum are both important quantities, and they are even more useful when used together. This section reveals that energy can similarly be useful when considering rotating extended objects.

While you Read the Section: Important Terms and Equations

Use the space below to define each term in your own words. For equations, use the space to identify what each letter represents and its associated SI unit. You may also add any other notes that will be helpful for future review.

rolling without slipping

$$K = K_{\text{translational}} + K_{\text{rotational}} = \frac{1}{2}Mv_{\text{CM}}^2 + \frac{1}{2}I_{\text{CM}}\omega^2$$

$$v_{\text{CM}} = R\omega$$

After you Read the Section: Check Your Understanding

Choose the best answer to each of the following. Use the space provided to write a short justification for your selection. When you're finished, check that you got the right answers for the right reasons!

1. Which of the following statements about translational kinetic energy and rotational kinetic energy is correct?

   a. Translational kinetic energy is conserved, but rotational kinetic energy is not conserved.

   b. Translational kinetic energy is not conserved, but rotational kinetic energy is conserved.

   c. Translational kinetic energy and rotational kinetic energy are each conserved, but separately from each other.

   d. Individually, translational kinetic energy and rotational kinetic energy are not conserved, but the total energy (including the sum of translational kinetic energy and rotational kinetic energy) is conserved.

   Justify your choice:

2. Which one of the following is an expression for total kinetic energy?

   a. $\frac{1}{2}mv^2$
   b. $\frac{1}{2}mv_{CM}^2 + \frac{1}{2}I_{CM}\omega^2$
   c. $\frac{1}{2}m(v+\omega)^2$
   d. $\frac{1}{2}I(v+\omega)^2$

   Justify your choice:

3. A wheel rolls without slipping. Its speed is v_0 and its angular speed is ω_0. If its speed increases to $2v_0$ and continues to roll without slipping, what will its new angular speed be?

   a. ω_0
   b. $2\omega_0$
   c. $4\omega_0$

   d. The wheel's angular speed is independent of its translational speed, so its new angular speed could have any value.

   Justify your choice:

4. A baseball is thrown through the air. Its speed is v_0 and its angular speed is ω_0. If it is thrown again with speed $2v_0$ what will its new angular speed be?

   a. ω_0
   b. $2\omega_0$
   c. $4\omega_0$

   d. The ball's angular speed is independent of its translational speed, so its new angular speed could have any value.

   Justify your choice:

11-3 Angular momentum is always conserved; it is constant when there is zero net torque

This section elaborates on how angular momentum is separate from, but analogous to linear momentum. It shouldn't be a surprise at this point to find that the equation for angular momentum $\left(L_z = I\omega_z\right)$ has the same form as the equation for linear momentum $\left(p_x = mv_x\right)$.

While you Read the Section: Important Terms and Equations

Use the space below to define each term in your own words. For equations, use the space to identify what each letter represents and its associated SI unit. You may also add any other notes that will be helpful for future review.

conservation of angular momentum

$L_z = I\omega_z$

$L = r_\perp p = rp\sin\phi$

After you Read the Section: Check Your Understanding

Choose the best answer to each of the following. Use the space provided to write a short justification for your selection. When you're finished, check that you got the right answers for the right reasons!

1. An object spins with an angular momentum of 8.0 kg·m²/s. If the object's shape changes resulting in its rotational inertia doubling, how much angular momentum will the object then have? Assume that no net torque was exerted on the object while its shape changed.

    a. 2.0 kg·m²/s   b. 4.0 kg·m²/s   c. 8.0 kg·m²/s   d. 16.0 kg·m²/s

    Justify your choice:

2. An object's angular momentum initially has magnitude L_0. If its rotational inertia and angular speed both double, how much angular momentum will it then have?

    a. $0.5 L_0$ b. L_0 c. $2 L_0$ d. $4 L_0$

    Justify your choice:

3. A 3.0 kg object's velocity is 5.0 m/s directed at 30° north of east when its position is $x = 4.0$ m, $y = 0$. What is the magnitude of the object's angular momentum with respect to a vertical axis that passes through $x = 0, y = 0$?

    a. 0   b. 7.5 kg·m²/s   c. 15 kg·m²/s   d. 30 kg·m²/s

    Justify your choice:

11-4 Newton's law of universal gravitation

This final section of the chapter introduces quite a few new terms, and connects back to universal gravitation, which was first introduced in Chapter 6 (force) and expanded on in Chapter 8 (energy). Reviewing these chapters, especially the sections that directly connect to orbital motion (Sections 6-4, 6-5, and 8-5), would be helpful before reading this section.

While you Read the Section: Important Terms and Equations

Use the space below to define each term in your own words. For equations, use the space to identify what each letter represents and its associated SI unit. You may also add any other notes that will be helpful for future review.

law of orbits

semimajor axis

Eccentricity

law of areas

law of periods

$$E = -\frac{Gm_{\text{Earth}}m}{2r}$$

$$T^2 = \frac{4\pi^2}{Gm_{\text{Sun}}}a^3$$

After you Read the Section: Check Your Understanding

Choose the best answer to each of the following. Use the space provided to write a short justification for your selection. When you're finished, check that you got the right answers for the right reasons!

1. Two planets have circular orbits around the same star. The planet closer to the star has orbital radius r_0 and moves with speed v_0. If the further planet has orbital radius $4r_0$, then what must its speed be equal to?

    a. $v_0/4$ b. $v_0/2$ c. $2v_0$ d. $4v_0$

    Justify your choice:

2. Two planets have circular orbits around the same star. The planet closer to the star has orbital radius r_0 and orbital period T_0. If the further planet has orbital radius $2r_0$, then what must its orbital period be equal to?

    a. $2\sqrt{2}T_0$ b. $\sqrt{2}T_0$ c. $T_0/2$ d. $T_0/(2\sqrt{2})$

    Justify your choice:

3. The total mechanical energy of an object in a circular orbit is -100 J. If the object's orbital radius increased, which of the following is a possible total mechanical energy of the object in its new orbit?

    a. -200 J   b. -100 J   c. -50 J   d. 100 J

    Justify your choice:

After you Read the Chapter: Test Yourself

After reading the chapter and trying the questions below, I recommend that you then **work on the Chapter 11 Review Problems** at the end of the chapter in your textbook before continuing into the next chapter.

End of Chapter Multiple Choice Questions

1. The diagram here shows an object moving with velocity v. Four points are marked A, B, C, and D. The object has the largest angular momentum with respect to an axis perpendicular to the plane of the diagram and passing through which of these four points?

    a. Point A b. Point B c. Point C d. Point D

For the next two questions: An astronaut is holding an object while on a space walk outside of the International Space Station. She and the object are initially at rest with respect to the station. She then throws the object. The object moves away from the station but is not spinning with respect to it.

2. Which of the following correctly describes the translational motion, if any, of the astronaut's center of mass after she throws the object?

    a. It will necessarily remain at rest with respect to the station.

    b. It will necessarily be moving towards the station.

    c. It may or may not move with respect to the station, depending on how massive the object is compared to her mass.

    d. It may or may not move with respect to the station, depending on how fast she threw the object.

3. Which of the following correctly describes the astronaut's rotational motion, if any, with respect to an axis of rotation passing through her center of mass after she throws the object?

    a. She will necessarily have no rotational motion about this axis.

    b. She will necessarily be rotating about this axis.

    c. She may or may not be rotating about this axis, depending on whether the object moves directly away from her center of mass.

    d. She may or may not be rotating about this axis, depending on the magnitude of the object's linear momentum.

4. Which one of the following diagrams best represents a possible orbit of a planet around a star (with the mass of the star being much greater than the mass of the planet)?

    a.   b.   c.   d.

Chapter 11 | Torque and Rotation II   163

5. Io is a moon that orbits Jupiter while Jupiter and Earth both orbit the Sun. Which of the following is a correct statement about these orbits?

   a. Io and Jupiter each sweep out the same area within their orbits in the same time interval.

   b. Earth and Jupiter each sweep out the same area within their orbits in the same time interval.

   c. The ratio of orbital radius cubed to orbital period squared is the same for Io and Earth.

   d. None of the above statements are true.

End of Chapter Problems

Solve each of the following problems on separate paper.

1. A 4.00 kg object moves in a line with a constant velocity of 6.00 m/s to the east. What is the magnitude of its angular momentum with respect to a vertical axis 3.00 m south of the object?

2. A circular platform is level and able to spin freely about its center. A spring-loaded launcher is attached to the platform so that it will launch a 15.0 g block at 4.00 m/s from a point that is 30.0 cm from the center of the platform as shown in the diagram here. The platform (including the launcher) has a rotational inertia of 0.0162 kg·m². If the platform and block were stationary before the block is launched, what will be the platform's angular speed immediately after launching the block?

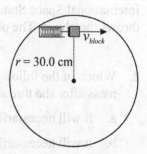

3. A solid disk has rotational inertia with respect to an axis through its center $I = \frac{1}{2}mr^2$. A solid disk with mass m and radius r that is free to spin around its center has a light string wrapped several times around its circumference. The string is then pulled, producing a constant tension F_T for a time t. If the disk was initially at rest, derive an expression for the angular speed it then has in terms of m, r, F_T and Δt.

4. Calculate the orbital radius of Mars in terms of Earth's orbital radius R_E, given that Mars has an orbital period of 687 days. Assume that both orbits are well-approximated to be circular.

Chapter 12
Oscillations Including Simple Harmonic

Before you Read the Chapter: Prepare Yourself

Be sure that you have a good understanding of:
- Ideal spring force (**Section 5-6**)
- Work done by a varying force (**Pages 311-314**)
- Spring potential energy (**Pages 321-322**)

Chapter Overview

Chapter 12 introduces a special type of motion known as simple harmonic motion. Springs and pendula make appearances in this chapter as both can be related to simple harmonic motion. Note that this chapter introduces many new equations, but a closer look reveals that many of these equations are built upon familiar ideas. In some cases, what may look like different equations are in fact different forms of the same equation, or combinations of previously introduced equations.

Learning Objectives

- ☐ Define oscillation and give everyday examples of systems that oscillate.
- ☐ Qualitatively and quantitatively explain characteristics of oscillations, including the key properties of simple harmonic motion, and what is required for simple harmonic motion to occur.
- ☐ Qualitatively describe a system using tensile stress and compressive stress and describe the relationship to Hooke's law.
- ☐ Explain the connection between Hooke's law and simple harmonic motion, and be able to calculate velocity and acceleration for any given displacement from equilibrium of an object oscillating on a spring.
- ☐ Explain what properties determine the period, frequency, and angular frequency of an object oscillating on a spring and how these characteristics of motion depend on those properties.
- ☐ Qualitatively and quantitatively predict how kinetic and potential energies vary during an oscillation of a system, and relate these changes to changes in the system's internal structure.
- ☐ Explain what properties determine the period, frequency, and angular frequency of a simple pendulum and how these characteristics of the motion depend on those properties.
- ☐ Explain what properties determine the period, frequency, and angular frequency of a physical pendulum and how these characteristics of the motion depend on those properties.

12-1 We live in a world of oscillations

This is a short section primarily introducing the main idea of the chapter—oscillations—along with 3 new terms that will be used regularly throughout the rest of the chapter.

While you Read the Section: Important Terms and Equations

Use the space below to define each term in your own words. You may also add any other notes that will be helpful for future review.

oscillation

cycle

period

After you Read the Section: Check Your Understanding

Choose the best answer to each of the following. Use the space provided to write a short justification for your selection. When you're finished, check that you got the right answers for the right reasons!

1. Which one of the following best describes the meaning of the word oscillation?

    a. One instance of a repeating pattern.

    b. A kind of motion in which an object moves back and forth around a position of equilibrium.

    c. The number of completed instances of a repeating pattern.

    d. The amount of time for each instance of a repeating pattern.

    Justify your choice:

2. Which one of the following best describes the meaning of the word cycle?

    a. One instance of a repeating pattern.

    b. A kind of motion in which an object moves back and forth around a position of equilibrium.

    c. The number of completed instances of a repeating pattern.

    d. The amount of time for each instance of a repeating pattern.

    Justify your choice:

3. Which one of the following best describes the meaning of the word period?

    a. One instance of a repeating pattern.

    b. A kind of motion in which an object moves back and forth around a position of equilibrium.

    c. The number of completed instances of a repeating pattern.

    d. The amount of time for each instance of a repeating pattern.

    Justify your choice:

12-2 Oscillations are caused by the interplay between a restoring force and inertia

This section introduces some new terms and reveals the relationship between period and frequency. Importantly, this section also reveals the precise force conditions needed to produce oscillating motion.

While you Read the Section: Important Terms and Equations

Use the space below to define each term in your own words. For equations, use the space to identify what each letter represents and its associated SI unit. You may also add any other notes that will be helpful for future review.

restoring force

frequency

hertz

amplitude

$f = \dfrac{1}{T}$ and $T = \dfrac{1}{f}$

After you Read the Section: Check Your Understanding

Choose the best answer to each of the following. Use the space provided to write a short justification for your selection. When you're finished, check that you got the right answers for the right reasons!

1. An object is dropped from a height of 1 m above the floor. It falls with negligible air resistance, collides elastically with the floor, and rises back up to a height of 1 m. It then falls, then bounces back up to a height of 1 m over and over. Is this motion an example of an oscillation?

   a. Yes, because the ball moves down and up in a regularly repeating way.

   b. Yes, because the impact with the floor provides a restoring force.

   c. No, because there is no equilibrium position at which there is no net force, and no restoring force pushing the object toward the equilibrium position.

   d. No, because the object moves vertically, and not along the *x*-axis.

   Justify your choice:

2. An oscillating object completes 5 cycles in 10 seconds. What is the period of the oscillation?

   a. 0.5 s   b. 0.5 Hz   c. 2 s   d. 2 Hz

   Justify your choice:

3. An oscillating object completes 5 cycles in 10 seconds. What is the frequency of the oscillation?

   a. 0.5 s   b. 0.5 Hz   c. 2 s   d. 2 Hz

   Justify your choice:

4. An object oscillates between $x = 2.0$ m and $x = 3.0$ m. What is the amplitude of the oscillation?

   a. 0.50 m   b. 1.0 m   c. 2.0 m   d. 3.0 m

   Justify your choice:

12-3 An object changes length when under tensile or compressive stress; Hooke's law is a special case

This section continues to lay the groundwork for the rest of the chapter. Only one "new" equation is introduced, though it is just a modified form of Hooke's law, which we have already worked with. Be sure to reach a good understanding of all three of these first "introductory" sections as the next section will be putting all of these ideas together!

While you Read the Section: Important Terms and Equations

Use the space below to define the term in your own words. Use the space to identify what each letter represents and its associated SI unit. You may also add any other notes that will be helpful for future review.

$F = k\Delta L$

After you Read the Section: Check Your Understanding

Choose the best answer to each of the following. Use the space provided to write a short justification for your selection. When you're finished, check that you got the right answers for the right reasons!

1. Which one of the following best describes the spring constant k?
   a. The length of the spring when no forces are exerted on it.
   b. The current length of the spring while it is stretched or compressed.
   c. A number representing the material the spring is made from.
   d. A number representing the stiffness of the spring.

   Justify your choice:

2. A student attaches one end of a spring to a support and then exerts a force F_1 to the other side. The student then measures the change in the spring's length from its relaxed length ΔL_1. The student repeats this for a range of different forces F_2, F_3, and F_4. The student then creates a graph with F on the vertical axis and ΔL on the horizontal axis. Which of the following correctly describes the resulting graph?
   a. A straight line with slope equal to spring constant k.
   b. A straight line with vertical intercept equal to the reciprocal of spring constant k.
   c. A straight line with vertical intercept equal to spring constant k.
   d. A parabolic curve with vertical intercept equal to spring constant k.

   Justify your choice:

12-4 The simplest form of oscillation occurs when the restoring force obeys Hooke's law

Be sure you have a good understanding of the first three introductory sections before continuing into this section. This section introduces many new terms and equations, so take your time reviewing. Also note that parts of this section are not tested on the AP® exam. Be sure to read the **AP® Exam Tip** boxes on Pages 566 and 570 for details on what is not tested from this section.

While you Read the Section: Important Terms and Equations

Use the space below to define each term in your own words. For equations, use the space to identify what each letter represents and its associated SI unit. You may also add any other notes that will be helpful for future review.

Hooke's law

angular frequency

harmonic property

simple harmonic motion (SHM)

sinusoidal function

phase angle

Chapter 12 | Oscillations Including Simple Harmonic

$F_{s,x} = -kx$

The simplest form of oscillation occurs when the restoring force obeys Hooke's law

$\omega = \sqrt{\dfrac{k}{m}}$

Be sure you have a good under standing of the first three introductory sections before continuing into this section. This section introduces many new terms and equations, so take your time reviewing. Also note that parts of this section are not tested on the AP® exam. Be sure to read the AP® Exam Tip box on Pages 566 and 570 for details on what is not tested from this section.

$T = \dfrac{2\pi}{\omega} = 2\pi\sqrt{\dfrac{m}{k}}$

While You Read the Section: Important Terms and Equations

Use the space below to define each term in your own words. For equations, use the space to identify what each letter represents and its associated SI unit. You may also add any other notes that may be useful for future review.

Hooke's law

$f = \dfrac{1}{T} = \dfrac{1}{2\pi}\sqrt{\dfrac{k}{m}} = \dfrac{\omega}{2\pi}$

angular frequency

$x = A\cos(\omega t + \phi)$

harmonic property

$v_x = -\omega A \sin(\omega t + \phi)$

simple harmonic motion (SHM)

sinusoidal function

$a_x = -\omega^2 A \cos(\omega t + \phi)$

phase angle

172   Section 12-4

After you Read the Section: Check Your Understanding

Choose the best answer to each of the following. Use the space provided to write a short justification for your selection. When you're finished, check that you got the right answers for the right reasons!

1. While moving in simple harmonic motion with amplitude A_0, the motion has angular frequency ω_0, period T_0, and frequency f_0. If the amplitude increases, which of the following would change?

    a. angular frequency
    b. period
    c. frequency
    d. none of these

    Justify your choice:

For the next 4 questions: An object is in equilibrium when its position is $x = 0$. It oscillates in simple harmonic motion with amplitude A along the x-axis.

2. At which one of the following positions does the object have an acceleration equal to zero?

    a. $x = -A$ b. $x = 0$ c. $x = 0.5A$ d. $x = A$

    Justify your choice:

3. At which one of the following positions does the object have an acceleration with the greatest magnitude directed in the $+x$ direction?

    a. $x = -A$ b. $x = 0$ c. $x = 0.5A$ d. $x = A$

    Justify your choice:

4. At which one of the following positions does the object have the greatest speed?

    a. $x = -A$ b. $x = 0$ c. $x = 0.5A$ d. $x = A$

    Justify your choice:

5. At which one of the following positions does the object have nonzero speed and nonzero acceleration?

    a. $x = -A$ b. $x = 0$ c. $x = 0.5A$ d. $x = A$

    Justify your choice:

12-5 Mechanical energy is constant in simple harmonic motion

With the foundation of simple harmonic motion having been laid out in the first four sections, this section applies energy and its conservation to it.

While you Read the Section: Important Terms and Equations

Use the space below to define each term in your own words. For equations, use the space to identify what each letter represents and its associated SI unit. You may also add any other notes that will be helpful for future review.

$$U_s = \frac{1}{2}kx^2$$

$$E = K + U_s = \frac{1}{2}kA^2$$

After you Read the Section: Check Your Understanding

Choose the best answer to each of the following. Use the space provided to write a short justification for your selection. When you're finished, check that you got the right answers for the right reasons!

For the next 2 questions: An object attached to a spring is in equilibrium when the object's position is $x = 0$, at which point the spring is relaxed. The object oscillates in simple harmonic motion with amplitude A along the x-axis.

1. At which one of the following positions does the object-spring system have maximum kinetic energy?

    a. $x = -A$ b. $x = 0$ c. $x = 0.5A$ d. $x = A$

    Justify your choice:

2. If the object-spring system has 10.0 joules of mechanical energy, then how much kinetic energy does the system have when the object is at $x = 0.5A$?

    a. 2.5 J   b. 5.0 J   c. 7.5 J   d. 10 J

    Justify your choice:

12-6 The motion of a pendulum is approximately simple harmonic

With all the needed aspects of simple harmonic motion having already been introduced, this section explains how within limits, the back-and-forth motion of a pendulum can often be well-approximated to be simple harmonic motion.

While you Read the Section: Important Terms and Equations

Use the space below to define each term in your own words. For equations, use the space to identify what each letter represents and its associated SI unit. You may also add any other notes that will be helpful for future review.

Pendula

simple pendulum

$$\omega = \sqrt{\frac{g}{L}}$$

$$T = \frac{2\pi}{\omega} = 2\pi\sqrt{\frac{L}{g}}$$

$$f = \frac{1}{T} = \frac{1}{2\pi}\sqrt{\frac{g}{L}} = \frac{\omega}{2\pi}$$

After you Read the Section: Check Your Understanding

Choose the best answer to each of the following. Use the space provided to write a short justification for your selection. When you're finished, check that you got the right answers for the right reasons!

1. Object 1 is attached to a spring and oscillates in simple harmonic motion. Object 2 is attached to a string and swings as a simple pendulum in simple harmonic motion. The motions of both objects have the same period $T = T_0$. If both objects are replaced with more massive objects which are then again set into simple harmonic motion, which of the following describes their motions?

    a. Both objects will continue to have the same period $T_{new} = T_0$.

    b. Both objects will again have the same period $T_{new} \neq T_0$.

    c. The object on the spring will continue to have the same period $T_{s,new} = T_0$, and the pendulum will have $T_{p,new} \neq T_0$.

    d. The pendulum will continue to have the same period $T_{p,new} = T_0$, and the object on the spring will have $T_{s,new} \neq T_0$.

    Justify your choice:

2. A pendulum swings with a frequency of 4 Hz. If the length of the pendulum is increased such that $L_{new} = 4L_{original}$, what will be its new frequency?

    a. 1 Hz   b. 2 Hz   c. 8 Hz   d. 16 Hz

    Justify your choice:

3. A small object is tied to a string and swings as a simple pendulum. Is the mechanical energy of the object constant throughout its motion?

    a. Yes, because mechanical energy is always constant for simple harmonic motion.

    b. Yes, because no external forces do work on the object throughout its motion.

    c. No, because the force of tension from the string is external and does work on the object.

    d. No, because the force of gravity from earth is external and does work on the object.

    Justify your choice:

12-7 A physical pendulum has its mass distributed over its volume

Simple pendula (introduced in the previous section) are idealized in that they are treated as if all of their mass is concentrated in a region with negligible volume. By comparison, a physical pendulum is one in which the mass is distributed. All pendula are physical pendula, but it is often acceptable to use the simple pendulum model (e.g., a "small object" attached to a "light string").

It is important to note that simple pendula are included on the AP® Physics 1 exam but physical pendulums (introduced in this section) are not, as explained in the AP® Exam Tip box on Page 588.

While you Read the Section: Important Terms and Equations

Use the space below to define each term in your own words. For equations, use the space to identify what each letter represents and its associated SI unit. You may also add any other notes that will be helpful for future review.

physical pendulum

$$\omega = \sqrt{\frac{mgh}{I}}$$

$$T = \frac{2\pi}{\omega} = 2\pi \sqrt{\frac{I}{mgh}}$$

$$f = \frac{1}{T} = \frac{1}{2\pi} \sqrt{\frac{mgh}{I}} = \frac{\omega}{2\pi}$$

After you Read the Section: Check Your Understanding

Choose the best answer to each of the following. Use the space provided to write a short justification for your selection. When you're finished, check that you got the right answers for the right reasons!

1. A uniform rod is supported so it can swing freely from one end as a physical pendulum. The rod is initially vertical. The lower end of the rod is then pulled to the side and released. As the rod swings back to its initial position, is the angular acceleration of the rod constant?

    a. No, because the torque provided by the gravitational force changes as its orientation changes.

    b. No, because the rotational inertia of the rod changes as its orientation changes.

    c. Yes, because there is no net torque exerted on the rod as its orientation changes.

    d. Yes, because the net torque exerted on the rod as its orientations remains constant.

    Justify your choice:

2. Increasing which one of the following variables would increase the period of a physical pendulum, provided that all other listed variables remained constant?

    a. The mass of the physical pendulum.

    b. The acceleration due to gravity.

    c. The rotational inertia of the physical pendulum about the pivot.

    d. The distance from the pivot point to the center of mass of the physical pendulum.

    Justify your choice:

After you Read the Chapter: Test Yourself

After reading the chapter and trying the questions below, I recommend that you then **work on the Chapter 12 Review Problems** at the end of the chapter in your textbook before continuing into the next chapter.

End of Chapter Multiple Choice Questions

For the next two questions: Objects 1 and 2 move in simple harmonic motion with amplitudes $A_2 = 2A_1$ and periods $T_2 = T_1$.

1. Which of the following correctly relates their average speeds v_1 and v_2?

   a. $v_2 = v_1$ b. $v_2 = \sqrt{2}\, v_1$ c. $v_2 = 2v_1$ d. $v_2 = 4v_1$

2. Which of the following correctly relates their angular frequencies ω_1 and ω_2?

   a. $\omega_2 = \omega_1$ b. $\omega_2 = \sqrt{2}\, \omega_1$ c. $\omega_2 = 2\omega_1$ d. $\omega_2 = 4\omega_1$

For the next two questions: A 4.0 kg object moves in simple harmonic motion along the *x*-axis. Its position *x* is given by the equation $x = 3.00 \cos(2.00t)$.

3. What is the period of the object's motion?

   a. 1.00 s   b. 2.00 s   c. 3.14 s   d. 6.00 s

4. What is the object's kinetic energy at time $t = 0$?

   a. 0   b. 3.14 J   c. 4.0 J   d. 8.0 J

5. The graph here shows the position of an object moving in Simple Harmonic Motion as a function of time. The object moves with speed v_1 at time $t = 1$ s, v_2 at time $t = 2$ s, v_3 at time $t = 3$ s and v_4 at time $t = 4$ s. Which of the following correctly ranks these speeds?

   a. $v_4 > v_3 = v_1 > v_2$
   b. $v_1 = v_3 > v_2 = v_4$
   c. $v_2 = v_4 > v_1 = v_3$
   d. $v_3 > v_2 = v_4 > v_1$

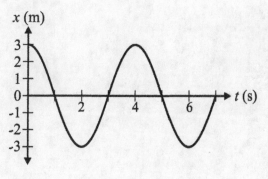

End of Chapter Problems

Solve each of the following problems on separate paper.

1. One end of a spring is attached to the ceiling of an elevator and the other end has a 2.5 kg object attached to it. While the elevator is accelerating upward at 3.0 m/s², the spring holds the object stationary with respect to the elevator, and is stretched 15 cm. What is the spring constant of this spring?

2. What is the length of a simple pendulum that runs a grandfather clock with a period of 2.00 s?

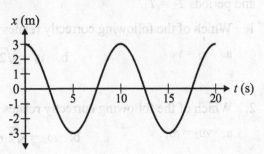

For the next 3 questions: The graph here shows the position of an object moving in SHM as a function of time.

3. What is the frequency of the oscillation?

4. What is the object's average speed over one full cycle?

5. What is the equation for the position of the object as a function of time?

Chapter 13
The Physics of Fluids

Before you Read the Chapter: Prepare Yourself

Be sure that you have a good understanding of:
- All previous chapters!

Chapter Overview

This is the final chapter of material in AP® Physics 1. All previous chapters considered various aspects of physical objects that were typically solid. This chapter is a significant departure as it introduces numerous aspects of "non-solids" (i.e., fluids). Note that this is an unusually long chapter consisting of ten sections and introduces many new vocabulary terms.

Learning Objectives

☐ Describe the similarities and differences between the properties of liquids and gases and how these relate to the interactions of their constituent molecules or atoms.

☐ Apply the definition of density to predict the densities, differences in densities, or changes in densities under different conditions for natural phenomena and design an investigation to verify the prediction.

☐ Explain the origin of fluid pressure in terms of molecular motion.

☐ Calculate the pressure at a given depth in a fluid in hydrostatic equilibrium.

☐ Explain the difference between absolute pressure and gauge pressure.

☐ Calculate the force on an object due to a difference in pressure on its sides.

☐ Explain how to apply Pascal's principle to a fluid at rest.

☐ Explain the source of the buoyant force on an object in a fluid, including how that determines its direction, and use Archimedes' principle to find that force.

☐ Use conservation of mass (the continuity equation) to analyze the flow of an incompressible fluid.

☐ Relate Bernoulli's principle to conservation of energy.

☐ Apply Bernoulli's equation to relate fluid pressure, height, and flow speed in an incompressible fluid.

☐ Use Bernoulli's equation and the continuity equation to describe the behavior of a moving ideal fluid.

☐ Describe the role of surface tension in the behavior of liquids.

13-1 Liquids and gases are both examples of fluids

The first section introduces the main topic of the chapter along with a few terms that will be important throughout the chapter. As there are nine more sections on fluids, be sure that you understand all new vocabulary terms before moving onto the next section!

While you Read the Section: Important Terms and Equations

Use the space below to define each term in your own words. You may also add any other notes that will be helpful for future review.

solid

fluid

liquid

gas

compressible fluid

incompressible fluid

After you Read the Section: Check Your Understanding

Choose the best answer to each of the following. Use the space provided to write a short justification for your selection. When you're finished, check that you got the right answers for the right reasons!

1. Which of the following states of matter are fluid?

    a. only gases   b. only liquids   c. only solids   d. gases and liquids

    Justify your choice:

2. Which one of the following is a property of all fluids?

    a. They maintain a nearly constant volume.

    b. They have a volume that is easily changed.

    c. Their molecules are in constant motion and can flow.

    d. Their molecules exert negligible forces on each other.

    Justify your choice:

3. Which one of the following best describes an ideal fluid?

    a. an incompressible liquid

    b. a compressible liquid

    c. an incompressible gas

    d. a compressible gas

    Justify your choice:

4. Which one of the following best describes the attractive forces between molecules in a fluid?

    a. They are strong in gases and weak in liquids.

    b. They are weak in gases and strong in liquids.

    c. They are strong in both gases and liquids.

    d. They are weak in both gases and liquids.

    Justify your choice:

13-2 Density measures the amount of mass per unit volume

This section is straightforward in that it introduces a concept (density) that is likely already familiar to you. With that said, be sure to have a clear understanding of what density is (and what it is not). Also remember that we will usually need density expressed in SI units for our purposes—that is generally not the case outside of physics courses.

While you Read the Section: Important Terms and Equations

Use the space below to define each term in your own words. For equations, use the space to identify what each letter represents and its associated SI unit. You may also add any other notes that will be helpful for future review.

density

specific gravity

$$\rho = \frac{m}{V}$$

After you Read the Section: Check Your Understanding

Choose the best answer to each of the following. Use the space provided to write a short justification for your selection. When you're finished, check that you got the right answers for the right reasons!

1. Which one of the following is the SI unit for density?

   a. kg/m^3 b. kg/L c. g/cm^3 d. g/L

   Justify your choice:

2. What is the mass of 50 mL of a substance that has a specific gravity of 2?

   a. 0.1 kg   b. 1 kg   c. 10 kg   d. 100 kg

   Justify your choice:

13-3 Pressure in a fluid is caused by the impact of molecules

Like the previous section, this section introduces one main idea (pressure) along with an equation to calculate it in addition to a couple of related terms and ideas. Pressure will be used extensively throughout the remainder of the chapter, so consider this to be a foundational idea that you must understand well!

While you Read the Section: Important Terms and Equations

Use the space below to define each term in your own words. For equations, use the space to identify what each letter represents and its associated SI unit. You may also add any other notes that will be helpful for future review.

pressure

pascal

atmosphere (unit of pressure)

$$P = \frac{F_\perp}{A}$$

Chapter 13 | The Physics of Fluids   185

After you Read the Section: Check Your Understanding

Choose the best answer to each of the following. Use the space provided to write a short justification for your selection. When you're finished, check that you got the right answers for the right reasons!

1. Three forces are exerted on an object. Force 1 has magnitude F_0 and is exerted over an area of A_0. Force 2 has magnitude $2F_0$ and is exerted over an area of A_0. Force 3 has magnitude $3F_0$ and is exerted over an area of $6A_0$. Which of the following correctly ranks the pressures associated with these forces?

    a. $P_1 > P_2 > P_3$ b. $P_3 > P_2 > P_1$ c. $P_2 > P_1 > P_3$ d. $P_2 > P_3 > P_1$

    Justify your choice:

2. Which of the following is closest to the force exerted by the atmosphere on a surface that is circular with a radius of 5 cm?

    a. 8×10^{-3} N   b. 8×10^1 N   c. 8×10^2 N   d. 8×10^6 N

    Justify your choice:

3. A gas in a container exerts pressure on the container walls. Which one of the following statements best describes the cause of this pressure?

    a. The molecules of gas are stationary; the molecules exert forces on the container in the same way that a compressed spring would because the gas is pressurized.

    b. The molecules of gas are stationary; the molecules exert forces on the container walls because their weight pushes them into the container walls.

    c. The molecules of the gas are stationary; the molecules exert forces on the container walls because the container walls exert force on the gas to keep them from moving.

    d. The molecules of the gas are in continuous motion; they exert forces on the container when they collide with it.

    Justify your choice:

13-4 In a fluid at rest pressure increases with increasing depth

This section reveals an important relationship between pressure and depth. The relationship is straightforward, but remember that this relationship has a limitation—the ideas and equations of this section are limited to ideal incompressible fluids that are stationary (i.e., not flowing) and in an equilibrium state (i.e., the pressure at any one point is constant). Pressure in flowing fluids is more complex and will be introduced later in the chapter.

While you Read the Section: Important Terms and Equations

Use the space below to define each term in your own words. For equations, use the space to identify what each letter represents and its associated SI unit. You may also add any other notes that will be helpful for future review.

hydrostatic equilibrium

uniform density

equation of hydrostatic equilibrium

millimeters of mercury

torr

gauge pressure

Chapter 13 | The Physics of Fluids

absolute pressure

$P = P_0 + \rho g d$

After you Read the Section: Check Your Understanding

Choose the best answer to each of the following. Use the space provided to write a short justification for your selection. When you're finished, check that you got the right answers for the right reasons!

1. A tall glass is half-filled with water. Which of the following would increase the pressure that the water at the bottom of the glass exerts onto the bottom of the glass?

    a. Adding more water to the glass.

    b. Adding an oil that floats on top of the water in the glass.

    c. Dipping your finger into the water in the glass so that your finger is partly submerged but touching only water.

    d. All of the above.

    Justify your choice:

2. Which of the following correctly relates the absolute pressure P_{abs}, gauge pressure P_{gauge}, and reference pressure P_{ref} that the gauge pressure is measured relative to?

    a. $P_{gauge} = P_{abs} + P_{ref}$

    b. $P_{gauge} = P_{abs} - P_{ref}$

    c. $P_{gauge} = (P_{abs})(P_{ref})$

    d. $P_{gauge} = (P_{abs}) / (P_{ref})$

    Justify your choice:

13-5 A difference in pressure on opposite sides of an object produces a net force on the object

This section extends the idea of pressure by considering the net force that can result from a difference in pressure from one side of an object to the opposite side. Although this is straightforward, it is important to note that the "net force" referenced here is the net force provided only by the fluid. For this to equate to the actual net force, there must either be no other forces exerted on the object, or all other forces exerted on the object must add to zero.

While you Read the Section: Important Terms and Equations

Use the space below to define the term in your own words. Use the space to identify what each letter represents and its associated SI unit. You may also add any other notes that will be helpful for future review.

$$F_{net} = (P_2 - P_1)A$$

After you Read the Section: Check Your Understanding

Choose the best answer to each of the following. Use the space provided to write a short justification for your selection. When you're finished, check that you got the right answers for the right reasons!

1. A window has an area of 2 m². Which of the following is nearest to the net force exerted on the window by the air if the pressure on one side of the window is 1 atm, and the pressure on the other side is 0.99 atm?

    a. 0.01 N   b. 0.02 N   c. 1000 N   d. 2000 N

    Justify your choice:

2. A block has pressure P_1 exerted on one side and a pressure $P_2 < P_1$ exerted on the opposite side. Which one of the following changes would necessarily increase the net force exerted on the block?

    a. Increasing both P_1 and P_2 by the same amount.
    b. Decreasing both P_1 and P_2 by the same amount.
    c. Doubling both P_1 and P_2.
    d. Increasing P_2 only.

    Justify your choice:

13-6 A pressure increase at one point in a fluid causes a pressure increase throughout the fluid

One useful application of fluid dynamics is in hydraulics: fluid pressure can be controlled at one end of a hose. The changes in pressure transmit through the hose and can exert forces at the other end. This section introduces this principle.

While you Read the Section: Important Terms and Equations

Use the space below to define the term in your own words. You may also add any other notes that will be helpful for future review.

Pascal's principle

After you Read the Section: Check Your Understanding

Choose the best answer to each of the following. Use the space provided to write a short justification for your selection. When you're finished, check that you got the right answers for the right reasons!

1. A plastic bottle is filled halfway with water and the cap is tightly secured. If the bottle is then squeezed to decrease its volume, which of the following correctly describes the changes in pressure, if any, that result in the bottle?

   a. The pressure at all points in the bottle will increase by the same amount.

   b. The pressure at all points in the bottle will remain constant.

   c. The pressure of the air in the bottle will increase but the pressure at points in the water will remain constant.

   d. The pressure at points in the water will increase but the pressure of the air in the bottle will remain constant.

   Justify your choice:

2. A hydraulic lift has area A_1 on one side and $A_2 = 2A_1$ on the other side. If a force F_1 is then exerted on the first side, and pushes the cap on that side downward a distance d_1, which of the following are correct expressions for the force F_2 exerted on the other side and the distance d_2 moved by the other cap?

   a. $F_2 = F_1$, $d_2 = d_1$

   b. $F_2 = 2F_1$, $d_2 = \frac{1}{2}d_1$

   c. $F_2 = \frac{1}{2}F_1$, $d_2 = 2d_1$

   d. $F_2 = 2F_1$, $d_2 = 2d_1$

   Justify your choice:

13-7 Archimedes' principle helps us understand buoyancy

This is a notable section as it introduces the buoyant force, which is the last type of force we will see in AP® Physics 1 (with gravitational force, normal force, friction force, and spring force being some examples of types of forces we have already worked with). Don't confuse normal force (which results from two solid surfaces interacting with each other) with buoyant force!

While you Read the Section: Important Terms and Equations

Use the space below to define each term in your own words. For equations, use the space to identify what each letter represents and its associated SI unit. You may also add any other notes that will be helpful for future review.

buoyant force

Archimedes' principle

$F_b = \rho_{fluid} V_{displaced} g$

After you Read the Section: Check Your Understanding

Choose the best answer to each of the following. Use the space provided to write a short justification for your selection. When you're finished, check that you got the right answers for the right reasons!

1. Object 1 is 1 kg and is a solid block made of low-density foam. Object 2 is 1 kg and is a solid block made of medium-density wood. Object 3 is 1 kg and is made of high-density metal in the shape of a boat. All 3 objects float when placed in water. How do the volumes of water displaced by each object compare?

   a. $V_1 = V_2 = V_3$ b. $V_1 < V_2 < V_3$ c. $V_1 > V_2 > V_3$ d. $V_1 = V_2 > V_3$

   Justify your choice:

2. A helium-filled balloon released in air floats up to the ceiling, but a water-filled balloon with the same volume drops down to the floor. Which of the following best explains this difference in behavior?

   a. Both balloons have an upward buoyant force but because helium is less dense than air, the helium balloon has a larger buoyant force than the water balloon.

   b. Helium is less dense than air and water is more dense than air, so the helium balloon has a buoyant force exerted on it, but the water balloon does not have a buoyant force exerted on it.

   c. Helium is less dense than air so the buoyant force exerted on it is greater than its weight whereas water is more dense than air, so its weight is greater than the buoyant force exerted on it.

   d. Water is more dense than helium so the water in the water-filled balloon has a greater pressure than the pressure of the helium in the helium-filled balloon.

   Justify your choice:

3. A 40 kg block of wood has a density of 800 kg/m³. The block is held fully submerged under water in a pool by a cable connecting the block to the pool. What is the tension in the cable?

   a. 100 N

   b. 400 N

   c. 500 N

   d. 900 N

   Justify your choice:

13-8 Fluids in motion: a more robust definition of an ideal fluid, and application of conservation of mass

This section introduces many new terms pertaining to flowing fluids. Note that some of these terms pertain to nonideal fluid flow, which is not tested on the AP Physics 1 exam, as indicated in the **AP® Exam Tip** box on page 639.

While you Read the Section: Important Terms and Equations

Use the space below to define each term in your own words. For equations, use the space to identify what each letter represents and its associated SI unit. You may also add any other notes that will be helpful for future review.

steady flow

unsteady flow

laminar flow

streamline

turbulent flow

viscosity

no-slip condition

boundary layer

inviscid flow

irrotational flow

continuity equation

volume flow rate

$A_1 v_1 = A_2 v_2$

After you Read the Section: Check Your Understanding

Choose the best answer to each of the following. Use the space provided to write a short justification for your selection. When you're finished, check that you got the right answers for the right reasons!

1. Which one of the following statements regarding the continuity equation applied to an ideal fluid flowing in a pipe is correct?
   a. The pressure of the fluid remains constant along the length of the pipe.
   b. The speed of the fluid remains constant along the length of the pipe.
   c. The pressure of the fluid is related to the cross-sectional area of the pipe.
   d. The speed of the fluid is related to the cross-sectional area of the pipe.

   Justify your choice:

2. The continuity equation $A_1 v_1 = A_2 v_2$ for an ideal fluid flowing is most directly based on which one of the following?
   a. the law of conservation of energy
   b. the law of conservation of momentum
   c. the law of conservation of angular momentum
   d. the law of conservation of mass

   Justify your choice:

3. When an ideal fluid flows from a pipe with a larger cross-sectional area to a pipe with a smaller cross-sectional area, which of the following correctly describes the fluid's speed and density?
   a. The fluid's speed and density both remain constant.
   b. The fluid's speed increases, and its density remains constant.
   c. The fluid's speed remains constant, and its density increases.
   d. The fluid's speed increases, and its density decreases.

   Justify your choice:

13-9 Bernoulli's equation, an expression of the work-energy theorem, helps us relate pressure and speed in fluid motion

This section introduces Bernoulli's principle, which to many people is counterintuitive. Fluid dynamics can be complicated, so it is important to realize that Bernoulli's principle and Bernoulli's equation are both based on idealized fluid flow, which is usually assumed in the context of AP® Physics 1. With that said, pay close attention to the limitations of Bernoulli's equation on pages 654-656.

While you Read the Section: Important Terms and Equations

Use the space below to define each term in your own words. For equations, use the space to identify what each letter represents and its associated SI unit. You may also add any other notes that will be helpful for future review.

Bernoulli's principle

Bernoulli's equation

$$P_1 + \frac{1}{2}\rho v_1^2 + \rho g y_1 = P_2 + \frac{1}{2}\rho v_2^2 + \rho g y_2$$

After you Read the Section: Check Your Understanding

Choose the best answer to each of the following. Use the space provided to write a short justification for your selection. When you're finished, check that you got the right answers for the right reasons!

1. Bernoulli's equation for an ideal fluid flowing is most directly based on which one of the following?

    a. the law of conservation of energy

    b. the law of conservation of momentum

    c. the law of conservation of angular momentum

    d. the law of conservation of mass

    Justify your choice:

2. Water flows in a pipe from point 1 to point 2. The cross-sectional areas of the pipe at points 1 and 2 are A_1 and $A_2 < A_1$. The heights of the pipe at points 1 and 2 are h_1 and $h_2 < h_1$. Which one of the following correctly compares the pressures at the two points?

    a. The pressure at point 2 is necessarily less than the pressure at point 1.

    b. The pressure at point 2 is necessarily equal to the pressure at point 1.

    c. The pressure at point 2 is necessarily greater than the pressure at point 1.

    d. The pressure at point 2 may be less than, equal to, or greater than the pressure at point 1.

    Justify your choice:

3. Water flows in a pipe from point 1 to point 2. The cross-sectional areas of the pipe at points 1 and 2 are A_1 and $A_2 < A_1$. The heights of the pipe at points 1 and 2 are h_1 and $h_2 > h_1$. Which one of the following correctly compares the pressures at the two points?

    a. The pressure at point 2 is necessarily less than the pressure at point 1.

    b. The pressure at point 2 is necessarily equal to the pressure at point 1.

    c. The pressure at point 2 is necessarily greater than the pressure at point 1.

    d. The pressure at point 2 may be less than, equal to, or greater than the pressure at point 1.

    Justify your choice:

13-10 Surface tension explains the shape of raindrops and how respiration is possible

Congratulations! You have made it to the last section of the last chapter that is tested on the AP® Physics 1 exam! However, the material introduced in this section is not tested on the AP® Physics 1 exam, as indicated in the **AP® Exam Tip** box on Page 656. With that said, do not skip this section—it is less than two pages long and includes an interesting introduction to the physics of surface tension.

While you Read the Section: Important Terms and Equations

Use the space below to define the term in your own words. You may also add any other notes that will be helpful for future review.

surface tension

After you Read the Section: Check Your Understanding

Choose the best answer to each of the following. Use the space provided to write a short justification for your selection. When you're finished, check that you got the right answers for the right reasons!

1. Which one of the following best explains why liquids tend to form spherical drops?

    a. The molecules have the maximum amount of energy when they form a spherical shape.

    b. The molecules have the minimum amount of energy when they form a spherical shape.

    c. For a given volume, a sphere has the largest surface area.

    d. For a given volume, a sphere has the smallest surface area.

    Justify your choice:

After you Read the Chapter: Test Yourself

After reading the chapter and trying the questions below, I recommend that you then **work on the Chapter 13 Review Problems** at the end of the chapter in your textbook before continuing into the next chapter.

End of Chapter Multiple Choice Questions

1. When calculating the net force exerted on an object by a fluid with the equation $F_{net} = (P_2 - P_1)A$ should the pressures P_1 and P_2 be absolute pressures or gauge pressures?

    a. Both pressures must be absolute pressures to correctly calculate the net force.

    b. Both pressures must be gauge pressures to correctly calculate the net force.

    c. The pressures may both be absolute pressures or may both be gauge pressures to correctly calculate the net force.

    d. Pressure P_2 must be absolute pressure and P_1 must be gauge pressure to correctly calculate the net force.

For the next four questions:

   a. If a fluid flows in a single pipe, the number of kilograms of fluid flowing per second must be the same everywhere along the length of the pipe.

   b. If the pressure at one point in a static fluid is increased, the pressure will increase by the same amount everywhere in the fluid.

   c. If an object is surrounded by a fluid, the fluid will exert an upward force on the object equal to the weight of the displaced fluid.

   d. The conservation of energy can be used to relate the speed, height, and pressure at two points along a streamline of a flowing fluid.

2. Which one of the above is best identified as Pascal's principle?

3. Which one of the above is best identified as the principle that the continuity equation is based on?

4. Which one of the above is best identified as the principle that Bernoulli's equation is based on?

5. Which one of the above is best identified as Archimedes' principle?

End of Chapter Problems

Solve each of the following problems on separate paper.

1. Liquid mercury has a density of 5430 kg/m³. If the pressure at the surface of the mercury is 1.0×10^5 Pa, how deep below the surface is the pressure equal to 2.0×10^5 Pa?

2. A pipe has a diameter of 20 cm. The pipe then branches out into two pipes, each with a diameter of 10 cm. If water flows in the wider pipe at 5 m/s, how fast will it flow in the narrower pipes? Assume that the water flows equally fast in both narrower pipes.

3. A 5.0 kg wood block has a density of 800.0 kg/m³. How much downward force must be exerted on the block to hold it stationary, completely submerged under water?

4. Water flows in a horizonal pipe. At one point the water has a pressure of 3.0×10^5 Pa and flows at 12 m/s. The pipe becomes narrower so that at a second point the water flows at 18 m/s. Calculate the pressure at the second point. Assume that the two points are at the same height.

Chapter 14
Preparing for the AP® Physics 1 Exam

Exam Overview

The exam is made up of two separate parts, each of which is worth 50% of the overall exam score:

Part 1: Multiple-Choice
- 1 hour and 30 minutes to complete 45-50 multiple choice questions.
- You may use a calculator.
- You are provided with a copy of the AP® Physics 1 Table of Information and the AP® Physics 1 Equations Sheet.
- There is no penalty for guessing or answering incorrectly.

After completing the Multiple-Choice section, you get a short break between the two parts.

Part 2: Free-Response
- You may use a calculator.
- You are provided with a copy of the AP® Physics 1 Table of Information and the AP® Physics 1 Equations Sheet.
- 1 hour and 30 minutes to complete 4 questions—one of each of the following:
    - **Mathematical Routines:** Use math (calculate numerical values and/or derive algebraic equations) to analyze and make predictions about a given scenario. This question may include additional components such as drawing a free-body diagram or sketching a graph. This question may also include components that require you to make a claim or prediction about the scenario and justify or explain the claim or prediction by applying one or more physics concepts.
    - **Translation Between Representations:** Connect different representations of a given scenario such as visual representations, derived equations, and drawn graphs. This question may also include components that require you to justify why your answers do or do not agree with each other, make and justify a prediction about another scenario, or explain how a change in the given scenario would affect your predictions.
    - **Experimental Design:** The "lab" question may require you to design a procedure that could be completed using equipment found in a typical high school science classroom to answer a question about a physical scenario and/or collect data that could then be analyzed. This question may also ask you to identify sources of experimental error.
    - **Qualitative / Quantitative Translation:** This question will test you on your ability to connect various aspects of a scenario, such as physical laws and/or equations. This question may also include components that require you to justify why your answers do or do not agree with each other, make and justify a prediction about another scenario, or explain how a change in the given scenario would affect your predictions.

Frequently Asked Questions

When are the AP® exams taken?

AP® exams are taken each year in early May. You can check the College Board's website for dates.

What if I'm sick on the day of the exam, or I know I won't be available on the scheduled exam day due to a conflict?

If you are sick on the exam day or if you have a conflict that prevents you from taking an AP® exam at its regularly scheduled time, you may be able to take a "Conflict Exam." More information about late testing can be found on the College Board's website if needed.

What do I need to bring with me to the exam?

You should bring:

- No. 2 pencils for your multiple-choice answer sheet (bring several sharpened pencils).
- A good quality eraser.
- Pens with black or dark blue ink for the free-response section. You may use a pencil in the free-response section if you prefer, but I recommend that you use a pen; if you make a mistake, you can cross it off much more quickly than you can erase it. Your work needs to be legible but does not need to be neat! Additionally, if you cross it out and later determine your work was correct, you can always circle it and ask that it be graded!
- A ruler or other straight edge.
- One or two calculators (see below for more on calculators).
- Government-issued or school-issued photo I.D.

What kind of calculator can I bring?

The College Board website details which calculators are and are not permitted, but generally you can bring a scientific calculator or even a graphing calculator, provided that the calculator does not have the ability to communicate wirelessly with other devices (e.g., Bluetooth) or have a QWERTY (i.e., "typewriter" style) keyboard. You can bring up to two calculators; if you have two available, you might as well have a backup on hand.

Do I have to clear my calculator's memory for the exam?

Somewhat surprisingly, no—you are allowed to have whatever you want stored in your calculator's memory. This means that if you want to (and if you have a calculator with the capability to store information), you can type notes or equations into your calculator before the exam, save them into the calculator's memory, and freely access them during the exam. This might sound like "cheating", but it is explicitly permitted.

With this fact in mind, I strongly discourage you from putting anything into your calculator's memory as you really need to know and understand physics to do well on the exam. Referring to notes or other material stored on your calculator during the exam would waste valuable time.

How will my AP® Physics 1 exam be scored?

During the exam you will record your answers to the multiple-choice questions on a "bubble sheet" that will be scanned and scored electronically. You will answer the free-response questions directly on the exam paper in the space that is provided for you. Hardworking, dedicated, actual human beings (AP® Physics teachers and college physics professors) will score your free-response section based on standardized scoring guides.

What information do I get after my exam has been scored?

Your "raw" actual exam score (i.e., the number of points you earned on the exam) is not released, and you will not find out how well you did on any specific question. Your exam score will be reported to you as a single digit from 1 to 5, with 5 being the highest score possible (which is why you should Strive for a 5!).

How is the final AP® exam score determined?

Although you will not know your raw score, it is used to determine the AP® exam mark that gets reported to you. Each year, several factors are considered to determine how raw scores are converted into the five-point scale, so it's not possible to know ahead of time exactly what percentage you need to get a 5, for example.

General Exam Tips

- Familiarize yourself with the AP® Physics 1 Table of Information and Equations Sheet. You can download them from the College Board's website. Go over all the equations summarized throughout Chapters 1 to 13 and be sure to memorize the additional formulas that are not provided. You could store them in your calculator's memory if needed, but remember that your time is better spent not looking up these things. The less you need to look things up, the better! Most importantly, ensure you know what the symbols in each equation stand for.

- Consider making yourself flash cards for the equations and facts that you need to memorize for the exam.

- Knowing equations and facts is not enough. You need to understand the concepts and be able to apply them to unfamiliar situations. You can best achieve this knowledge with practice. In addition to everything in this workbook, you should practice with the questions and problems in the textbook, as well as with actual College Board questions, which are available on their website.

- There are no "trick questions" on the exam. My own definition of a "trick question" is a question that is designed to deceive you. None of the questions on the AP® exam are intended to deceive anyone, but they are designed to test you! Many questions are designed in such a way that a student with an incomplete understanding will likely get the question wrong. Again, these are not "trick questions," but rather what I would call good questions.

- Most questions are designed to test a specific topic or concept – try to identify what is being tested, as doing so can help you to focus your solutions.

- Some questions are designed to test your ability to connect two or more different topics – try to identify all topics that are relevant to these questions to focus your attempt to answer these questions.

- It's impossible to completely avoid getting stressed about the exam, but try to avoid being overly anxious before and during the test. If you think that your anxiety may interfere with your ability to do your best, consider talking to your teacher or guidance counselor about how to best manage test anxiety.

- Get as much practice with authentic AP® Physics 1 style questions as you can. In addition to the two full practice exams in this book, your textbook has two additional practice exams in it. You should do all four of these. Many of the textbook end of chapter problems emphasize skills needed on the exam, and each chapter ends with AP-style questions written by people who have authored AP® Physics 1 exam questions. You can download actual AP® Physics 1 free-response exam questions from previous years directly from the College Board's website and do them as well!

- Get everything that you will need for the exam ready to go no later than the day before the exam (see "What do I need to bring with me to the exam?" earlier in this section).

- You may be tempted to cram all the information the night before the exam, but this is not the best way to prepare for it! Aim to get most of your preparation done well before the exam so that you need to do only one last small review the night before. Then get a good night's sleep.

- Eat a good breakfast and be sure to arrive on time!

- Listen very carefully to all instructions given at the exam and be sure to follow them closely.

Multiple Choice Tips

- Expect a wide range of difficulty in the multiple-choice section. There will be some very straight-forward questions, some very challenging questions, and some questions between these extremes.

- The questions are generally not ordered by topic, nor by difficulty. Don't let the lack of organization throw you off. It is possible you might encounter some challenging questions early on. Don't let them discourage you!

- Time is of the essence! You have only 1.8 minutes on average for each question. If you think it will take more than 3 minutes to answer a question, circle it and move on to the next question. Come back to tackle the harder questions after you have completed all the easier ones if time remains to do so.

- Don't leave any questions unanswered, as there is no penalty for getting a question wrong. If you are unable to finish all the questions within the given time, use the last few minutes to bubble in guesses for the ones you didn't answer. You will not be given extra time to do this, so be sure to bubble in an answer for every question before the time is up.

- Read each question carefully. It is often the case that a single missed or misread word can change the nature of the question.

- Don't answer questions based only on your intuition. Your intuition might be correct, but always answer the questions based on correct physics.

- Be suspicious of any question that appears to have an "obvious" answer! The obvious answer may be correct, so don't rule it out, but think carefully about each question to avoid deciding on an answer too quickly.

- If you can't figure out the correct answer to a question, you may still be able to eliminate one or two of the listed options. Each correctly eliminated answer increases your odds of getting the question right, even if you need to guess from the remaining options.

- You are not permitted to use scrap paper, but you can write directly onto the question booklet (nothing you write in it in this section will be graded). It can be helpful to underline key words within questions, or to use the margins for scratch work.

- You can also write on your formula sheet, but you should be aware that you will be handing it in at the end of the multiple-choice section and given a new copy for the free-response section. If you want to add anything to your formula sheet (from memory) so that you can visually refer to it during the exam, you will have to make your additions in both halves of the exam.

Free Response Tips

- You don't have to answer these in the same order they appear on the exam! Start by quickly looking over each of the four questions. Start with the question that you feel most comfortable with and save the one you are least comfortable with for last.
- Don't spend too long on any one question if that time is better spent answering other parts of the exam. As soon as you realize that you are really stuck on a question, move on, and come back later if you have time.
- Your work doesn't have to be overly neat, but it needs to be legible.
- If you are unsure how to do an earlier part of a question but the answer is needed for a later part in the question, you can make up an answer for the earlier part to proceed with the later part. You will not get credit in the earlier part, but you may get full credit in the later parts if the work you show is correct.
- Answer the question! This might sound obvious, but students often fail to identify exactly what is being asked in a question. For example, if the question is "Does the object move with constant velocity?" don't answer with an equation – answer with "yes" or "no." Make your answer easy to find by drawing a box around it.
- Don't go beyond what was asked for! Answer the question completely and correctly, and then move on. You will not get extra credit for doing extra work, but you would be wasting time, and you may get penalized if your extra work contains errors.
- Look for the following key words, each of which has a specific meaning:
    - If you are asked to **calculate** something, it means that you must show your work, beginning with general physics formulas or principles. Your answer will generally be numeric; don't forget to include the correct unit for your answer!
    - If you are asked to **derive** something, then you generally need to start with one or more fundamental equations, such as those provided on the equations sheet. You will then need to algebraically work with the equation(s) to derive what you were asked to derive, usually in terms of a prescribed list of variables or symbols for specific quantities. Be sure to use the exact same symbols that the question has provided to you in these cases! If the question identifies the acceleration as a_1, for example, your answer must not merely reference a generic acceleration a.
    - If you are asked to **determine** something (or if the question asks "**what is ...**"), then you do not need to show supporting work, though it would still be a good idea to show work if it is needed. Questions usually ask you to "determine" something if the answer can be found with minimal work.
    - If you are asked to **sketch** a graph, you generally do not need specific points, but rather you should draw a graph that shows key features that would be expected on this graph. A sketch of a graph should correctly show straight lines, curves, intercepts, and asymptotes as appropriate.
    - If you are asked to **plot** data, you will need to show each point for data that has been provided for you or which you have calculated earlier in the question. This may be on a grid that has been provided, or you may need to make the grid yourself. Be sure to label axes, including units, if needed.

- If you are asked to **draw a line of best fit** based on plotted points, use a straight edge to draw one single straight line that shows the trend suggested by the points. If you then need to calculate a slope or determine an intercept on the graph, base them on the line of best fit, and not on the plotted points.
- If you are asked to **draw a free-body diagram** (or a force diagram), follow the specific directions that will be included. These directions usually indicate that you must not draw force components, and each arrow representing a force must start on the object and point away from it in the direction of the force.

Tips from Actual Students who took the AP® Physics Exam

The following are tips from my own students, given to me for inclusion here shortly after they took their AP® Physics 1 exams:

- Get a good scientific calculator that has all the necessary features but is also simple to operate. Use it for all your work before the exam so that you really know how to use it without even needing to think about it.
- Don't get discouraged when you get a practice problem wrong. Mistakes provide a perfect opportunity for you to fill in gaps in your understanding! Identify the concept that you didn't understand and work to establish a solid foundation on that concept. This way, you will be able to solve problems you have never even seen before (this will happen on the exam) purely based on your understanding of the subject at a conceptual level.
- Stop memorizing specific methods used to solve specific problems. Studying for AP® Physics 1 should be a process of gradually attaining a profound conceptual understanding rather than blindly memorizing facts, figures, and formulas.
- Aim to understand the context of equations rather than simply memorizing them.
- Pay attention to what is and is not in the system being considered. This is especially important when considering things like work and energy, and momentum and impulse.
- Use your time wisely on the multiple-choice questions. You will wish you had more time!
- Make sure that you deeply understand how to work with graphs (slope, area, intercepts, and equations for lines), including graphs that you've never seen before.
- Use the break time to rest!
- In addition to redoing problems from past problem sets/unit tests that I got wrong, I made a sheet for every topic and wrote down concepts that I tended to forget. Something can be as simple as "the coefficient of static friction can be greater than 1." These sheets were my personalized "ultimate study guide" and helped to ensure that I focused on the things I needed to review. When doing my final review, I focused on these sheets for a final refresh.
- Read the scoring guides from past AP® exams (they are available online). In addition to helping, you learn what you need to include in your answers, the scoring guides can also help boost your confidence by showing you how easy it can be to earn some of the points.
- If you don't know what to study, ask yourself "which topic will I mess up if I get a question on it?" and then you'll know.
- Be confident going into your exam. You can do it!

Practice Exam I

Section 1 – Multiple Choice Questions

Time – 1 hour and 30 minutes
50 Questions

Note: To simplify calculations, you may use $g = 10$ m/s^2 in all problems.

Directions: For each of the following questions, select the best answer from the four options listed.

1. A small ball is thrown upward. At $t = 0$ it is at position $y = 0$ and its speed is 20 m/s. Which of the following statements correctly describes the ball's position y and speed v at $t = 4$ s?

    a. $y = 0$; $v = 20$ m/s
    b. $y = 80$ m; $v = 20$ m/s
    c. $y = 80$ m; $v = 60$ m/s
    d. $y = 0$; $v = 60$ m/s

2. A 5 kg cart rolls at 3 m/s toward a wall. The cart collides with the wall and bounces off of it, moving at 1 m/s away from the wall. What is the magnitude of the change in the cart's momentum?

    a. 5 kg·m/s
    b. 10 kg·m/s
    c. 15 kg·m/s
    d. 20 kg·m/s

3. Four small objects, each with mass m, are positioned one each at the four corners of a square with side length L. What is the rotational inertia of the system composed of the four small objects with respect to an axis that passes through the center of the square?

    a. mL^2
    b. $2mL^2$
    c. $3mL^2$
    d. $4mL^2$

4. An object moves 100.0 m with a constant speed of 2.5 m/s. Its speed then changes to 10.0 m/s in a negligible amount of time. It then continues to move an additional 100.0 m with a constant speed of 10.0 m/s. What was the object's average speed?

   a. 4.0 m/s

   b. 6.25 m/s

   c. 8.5 m/s

   d. 12.5 m/s

5. A 5.00 kg object is on the floor of an elevator that is moving upward while slowing at a rate of 2.00 m/s². The magnitude of the normal force exerted on the object by the elevator floor is most nearly

   a. 10.0 N

   b. 40.0 N

   c. 50.0 N

   d. 60.0 N

Questions 6-7 refer to the following material:

Two projectiles are launched from the ground with the same initial speed v. The first projectile is launched at an angle of 30.0° above the horizontal, reaches a maximum height h_1 and lands on the ground a distance d_1 from its launch point. The second projectile is launched at an angle of 60.0° above the horizontal, reaches a maximum height h_2 and lands on lands on the ground a displacement d_2 from its launch point.

6. Which of the following correctly ranks the heights h that the projectiles reach and their horizontal displacements Δx?

   a. $h_1 = h_2$; $\Delta x_1 = \Delta x_2$

   b. $h_1 < h_2$; $\Delta x_1 > \Delta x_2$

   c. $h_1 = h_2$; $\Delta x_1 > \Delta x_2$

   d. $h_1 < h_2$; $\Delta x_1 = \Delta x_2$

7. Which of the following correctly compares the average velocities of the projectiles over the intervals beginning at the time each is launched until the time each land on the ground?

   a. $v_{average, 1} > v_{average, 2}$

   b. $v_{average, 1} < v_{average, 2}$

   c. $v_{average, 1} = v_{average, 2} > 0$

   d. The comparison of $v_{average, 1}$ and $v_{average, 2}$ depends on the magnitude of the launch speed v.

8. Planet 2 has a density that is half that of planet 1 and a radius that is four times that of planet 1. Which of the following is the acceleration due to gravity near planet 2's surface in terms of the acceleration due to gravity near planet 1's surface, g_1?

    a. $g_1/32$

    b. $g_1/16$

    c. $g_1/8$

    d. $2g_1$

9. A small ball is placed on top of a compressed spring which has its lower end on the ground and its upper end initially held in place. The upper end of the spring is then released. The spring extends upwards and pushes the ball. Which of the following indicates the changes, if any, to the mechanical energy of each of the following systems while the ball is rising with increasing speed, and is still in contact with the spring?

	Ball-Spring-Earth System	Ball-Spring System	Ball
a.	decreasing	increasing	decreasing
b.	constant	decreasing	increasing
c.	decreasing	decreasing	increasing
d.	constant	constant	constant

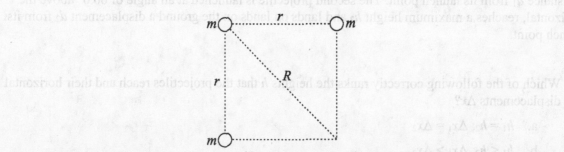

10. As shown in the figure above, three spheres with mass m are fixed at the corners of a square with side length r and diagonal length R. A fourth sphere also with mass m is initially far away from the other spheres. The system of four spheres has potential energy U_1. The fourth sphere is then moved to the empty corner of the square after which the system of four spheres has potential energy U_2. What is the change in the gravitational potential energy $U_2 - U_1$ of the four-sphere system?

    a. $-Gm^2\left(\dfrac{2}{r} + \dfrac{1}{R}\right)$

    b. $\dfrac{-3Gm^2}{r+R}$

    c. $\dfrac{-3Gm^2}{rR}$

    d. $\dfrac{-Gm^2}{\sqrt{r^2 + R^2}}$

11. Which of the following best describes the linear momentum of a satellite moving in a circular orbit around Earth with constant speed?

    a. The linear momentum of the satellite is constant because it moves with constant speed.

    b. The linear momentum of the satellite is constant because the net force on the satellite is perpendicular to its velocity.

    c. The linear momentum of the satellite is not constant because Earth's gravity does work on the satellite.

    d. The linear momentum of the satellite is not constant because the net force exerted on it is not zero.

12. Which of the following best describes the angular momentum of a satellite moving in a circular orbit around Earth with constant speed?

    a. The angular momentum of the satellite is constant because it moves with constant angular speed and has constant rotational inertia.

    b. The angular momentum of the satellite is constant because there is no net force on the satellite.

    c. The angular momentum of the satellite is not constant because the direction of the satellite's velocity is not constant.

    d. The angular momentum of the satellite is not constant because Earth's gravity exerts a torque on the satellite.

13. Can an object move with constant velocity if only forces with equal magnitudes are exerted on it?

    a. Yes, but only if there is an even number of forces.

    b. Yes, for any number of forces greater than one.

    c. No; if forces are applied to the object, it must accelerate.

    d. No; an object could only move with constant velocity if the forces had different magnitudes.

14. Bernoulli's equation is based on which law of Physics?

    a. conservation of linear momentum

    b. conservation of angular momentum

    c. conservation of energy

    d. Newton's first law

15. A block is placed on a surface that is inclined. The block remains at rest due to a static force of friction exerted on the block by the surface. If the angle of the incline is then increased, but the block continues to remain at rest on the surface, which of the following describes the change, if any, to the coefficient of static friction μ_S and the magnitude of the static friction force F_S exerted on the block?

    a. μ_S will increase and F_S will remain unchanged.

    b. μ_S will remain unchanged and F_S will increase.

    c. μ_S and F_S will both increase.

    d. μ_S and F_S will both remain unchanged.

16. An object is launched upward with speed v_0 from the surface of a small moon that has radius r. The object rises until it reaches a maximum height of r above the moon's surface and then falls back to the surface. The object is then launched upward from the moon's surface again but with an initial speed of $2v_0$. The greatest distance from the center of the moon that the object then reaches is

    a. $3r$

    b. $4r$

    c. $5r$

    d. The object will never stop moving away from the moon.

17. A container of water is on the floor. The volume flow rate of the water from a small hole in the side of the container depends on all of the following except for which one of the following?

    a. The distance to the bottom of the container below the hole.

    b. The height of the liquid above the hole.

    c. The area of the hole.

    d. The acceleration due to gravity.

18. A container of water is on the floor. Water flows horizontally out of a small hole in the side of the container and then lands on the ground. The horizontal displacement of the water as it flows from the container to the ground depends on all of the following except for which one of the following?

    a. The height of the hole above the bottom of the container.

    b. The height of the liquid above the hole.

    c. The area of the hole.

    d. The strength of the gravitational field.

19. A student starts a pendulum swinging with a small amplitude and carefully times each cycle. The student notices that the time for the pendulum to complete each cycle gradually increases. Which of the following could account for this observation?

    a. The length of the pendulum is steadily increasing.

    b. The length of the pendulum is steadily decreasing.

    c. The amplitude of the pendulum is steadily increasing.

    d. The amplitude of the pendulum is steadily decreasing.

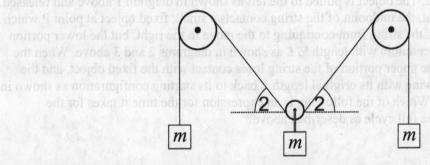

20. Three objects of equal mass m and three ideal pulleys are arranged as shown above. The two larger pulleys are free to turn but are fixed in position. The smaller pulley is free to turn but is supported only by the light string on top of which it rests. At what angle θ above the horizontal must each slanted section of string be oriented for the three objects to remain at rest?

    a. 30.0°

    b. 45.0°

    c. 60.0°

    d. It is not possible for the three objects shown above to be in equilibrium.

21. A block hangs in equilibrium from a vertical spring near Earth's surface. The block is then pulled down an additional 5 cm at which point the total length of the spring is 55 cm. The block is released, resulting in it oscillating in simple harmonic motion with an amplitude of 5 cm and period T_E. The same block and spring are relocated to the surface of the Moon where $g_{Moon} < g_{Earth}$. The spring is again pulled to a total length of 55 cm and the block is released. The block oscillates in simple harmonic motion. Which of the following correctly describes the amplitude A and period T_M of the block's motion on the Moon?

    a. $A = 5$ cm; $T_M = T_E$

    b. $A > 5$ cm; $T_M > T_E$

    c. $A = 5$ cm; $T_M > T_E$

    d. $A > 5$ cm; $T_M = T_E$

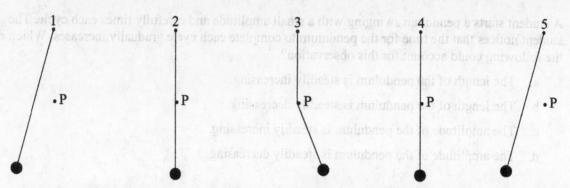

22. A pendulum with length L is constructed by fixing a small object to the end of a light string that is supported at its upper end. The object is pulled to the left as shown in diagram 1 above and released. When the string is vertical, the midpoint of the string contacts a small, fixed object at point P which prevents the upper half of the string from continuing to the move to the right, but the lower portion continues to swing as a pendulum with length ½ L as shown in diagrams 2 and 3 above. When the pendulum swings back the upper portion of the string loses contact with the fixed object, and the pendulum continues to swing with its original length L back to its starting configuration as shown in diagrams 4 and 5 above. Which of the following is an expression for the time it takes for the pendulum to complete one full cycle as described above?

   a. $2\pi\sqrt{\dfrac{L}{g}}$

   b. $2\pi\sqrt{\dfrac{L}{2g}}$

   c. $\pi\sqrt{\dfrac{L}{g}}\left(1+\dfrac{1}{\sqrt{2}}\right)$

   d. $\pi\sqrt{\dfrac{L}{g}}\left(1+\sqrt{2}\right)$

23. Balsa wood has a density of 130 kg/m³. If a block of balsa wood floats on water (which has a density of 1000 kg/m³), what percentage of the wood floats above the surface of the water?

   a. 87%
   b. 75%
   c. 25%
   d. 13%

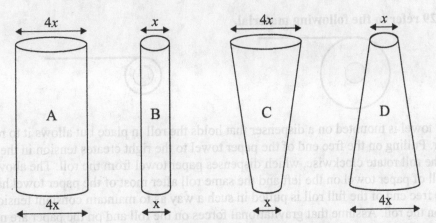

24. Four different open containers with equal heights but different shapes are filled with water as shown above. Which of the following correctly ranks the pressure in the water at the bottom of the containers?

   a. $P_C < P_A = P_B = < P_D$
   b. $P_D < P_A = P_B = < P_C$
   c. $P_B < P_C = P_D < P_A$
   d. $P_A = P_B = P_C = P_D$

25. A 2 kg object is attached to the lower end of a relaxed spring with a spring constant of 400 N/m. The object is then dropped. How far does the object move downward before it reaches the point of momentarily having a velocity of zero?

   a. 5 cm
   b. 10 cm
   c. 5 mm
   d. 10 mm

26. Two objects have mass $m_2 > m_1$. Each begins with zero velocity on a level surface that exerts negligible friction on it. Both objects then have equal horizontal net forces exerted on them over the same distance, at which point the net force is removed. How do the magnitudes of the momenta of the objects compare after the net force has been removed from both?

   a. $p_1 = p_2 = 0$
   b. $p_1 = p_2 > 0$
   c. $p_1 > p_2 > 0$
   d. $p_2 > p_1 > 0$

Questions 27-29 refer to the following material.

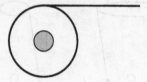 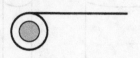

A roll of paper towel is mounted on a dispenser that holds the roll in place but allows it to rotate freely about its center. Pulling on the free end of the paper towel to the right creates tension in the paper towel which makes the roll rotate clockwise, which dispenses paper towel from the roll. The above diagram shows a full roll of paper towel on the left and the same roll after most of the paper towel has been dispensed. The free end of the full roll is pulled in such a way as to maintain constant tension in the paper towel pulling on the roll. Assume that gravitational forces on the roll and on the paper are negligible.

27. Which of the following best describes the magnitude of the torque provided to the roll by the tension force as the size of the roll decreases?

    a. It will be increasing.

    b. It will be decreasing.

    c. It will remain constant.

    d. It will initially increase and then decrease as the size of the roll decreases.

28. Which of the following best describes the rotation of the roll as the size of the roll decreases?

    a. The magnitude of its angular velocity will remain constant.

    b. The magnitude of its angular velocity will increase at a constant rate.

    c. The magnitude of its angular acceleration will increase.

    d. The magnitude of its angular acceleration will decrease.

29. Which of the following best describes the magnitude of the angular momentum of the system composed of all the paper (both the paper still on the roll, as well as the paper that has been dispensed), with respect to an axis that passes through the center of the roll?

    a. It will increase as the size of the roll decreases.

    b. It will decrease as the size of the roll decreases.

    c. It will remain constant as the size of the roll decreases.

    d. It will initially increase and then decrease as the size of the roll decreases.

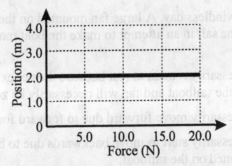

30. The figure above shows the position of an object as a function of a force exerted on the object. The object starts at rest at position $x = 2.0$ m. The force is introduced slowly, increasing in magnitude from 0 N to 20.0 N at a constant rate. What is the work done by this force on the object?

    a. 0
    b. 20.0 J
    c. 40.0 J
    d. 60.0 J

Questions 31-32 refer to the following material.

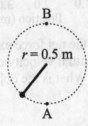

The figure above shows a 0.500 kg ball attached to a light stiff rod that is made to move in a vertical circle with radius $r = 0.500$ m. Point A is the lowest point on the circle and point B is the highest point. The ball begins at point A with speed $v_A = 1.00$ m/s and moves along the circular path to point B where its speed is $v_B = 2.00$ m/s.

31. How much work did the rod do to the Earth-ball system?

    a. 0
    b. 1.0 J
    c. 5.00 J
    d. 5.75 J

32. What is the magnitude of the vertical component of force that the rod exerts on the ball when it is at point B?

    a. 1.00 N
    b. 4.00 N
    c. 5.00 N
    d. 9.00 N

33. A sailboat floats at rest on a windless day. A large fan mounted on the back of the boat is then operated to blow air toward the sail in an attempt to make the boat move. Which of the following best describes the outcome?

   a. The sailboat will necessarily remain at rest because the net external force exerted on the system composed of the sailboat and fan will necessarily be zero.

   b. The sailboat will necessarily move forward due to forward force the air exerts on the sail.

   c. The sailboat will necessarily start to move backwards due to backward force the air exerts on the fan which is mounted on the sailboat.

   d. The sailboat may move forward, backward, or not at all, depending on the overall motion, if any, given to the air by the system composed of the sailboat and fan.

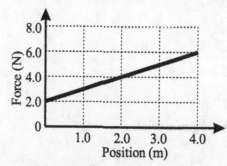

34. The figure above shows the magnitude of the net force exerted on a 2.0 kg object as a function of position. The object was moving with an initial velocity of 5.0 m/s at position $x = 0$, and the net force is in the direction opposite of its velocity. What is the object's kinetic energy when its position is 4.0 m?

   a. 9.0 J
   b. 16 J
   c. 25 J
   d. 41 J

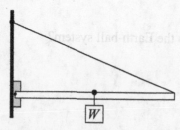

35. The figure above shows a horizontal uniform pole which is free to rotate about a hinge at its left end. A weight W hangs from the pole at the midpoint of the length of the pole. A light rope attached at the right end of the pole connects to a support. If the rope breaks, what will be the magnitude of the initial translational acceleration a of the weight W?

   a. $a = 0$
   b. $0 < a < g$
   c. $a = g$
   d. $a > g$

Questions 36 and 37 Refer to the following material.

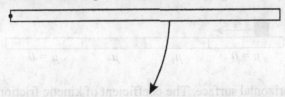

A rod at rest on a horizontal table with negligible friction is free to rotate about one end. The rod is initially at rest and oriented with the length of the rod to the right of the pinned end as shown above. It then begins to uniformly accelerate clockwise at $2\pi/3$ rad/s^2.

36. Which of the following best describes the orientation of the rod 3 seconds later?

    a. The length of the rod is to the right of the pinned end.

    b. The length of the rod is to the left of the pinned end.

    c. The length of the rod is above the pinned end, perpendicular to the rod's initial position.

    d. he length of the rod below the pinned end, perpendicular to the rod's initial position.

37. If the rotational inertia of the rod around the pinned end is 3.0 kg·m^2, then what is the rotational kinetic energy of the rod 3.0 seconds after it starts to turn?

    a. $2\pi^2$ J

    b. $4\pi^2$ J

    c. $6\pi^2$ J

    d. $8\pi^2$ J

38. A 2.0 kg object moving east at 3.0 m/s collides with a 10.0 kg object moving west at 3.0 m/s. If the two objects stick together after the collision, what will be their velocity immediately after the collision?

    a. 0

    b. 2.0 m/s west

    c. 2.0 m/s east

    d. 3.0 m/s west

Questions 39-40 refer to the following material.

A small object slides on a horizontal surface. The coefficient of kinetic friction between the surface and the object is different for different sections of the surface as shown here, with $\mu_2 > \mu_1 > 0$. The object initially slides on a section of the surface with negligible friction at constant speed v_0. It slides across the next two sections of the surface, each of which exerts a friction force onto the object. The object then slides with final constant speed $v_1 < v_0$ across the final section of the surface which exerts negligible friction on the object. Assume that the object's transition from one section to the next section occurs in a negligible amount of time.

39. Which of the following velocity-time graphs best represents the motion of the object as it slides across the surface?

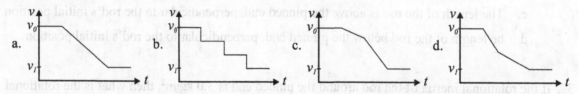

40. If only the order of the two sections with μ_1 and μ_2 were reversed and the object were sent sliding with the same initial speed v_0 as before, how would the new final speed v_2 compare to the original final constant speed v_1?

   a. $v_2 > v_1$

   b. $v_2 = v_1$

   c. $0 < v_2 < v_1$

   d. It is possible that the object will not make it through this time.

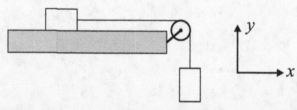

41. Two blocks are tied together with a light string which passes over an ideal pulley as shown above. There is negligible friction exerted by the surface on the block which is on it. Which of the following describes the initial motion, if any, of the center of mass for the system composed of the two blocks after they are released from rest?

   a. The system's center of mass remains stationary.

   b. The system's center of mass accelerates only in the +x direction.

   c. The system's center of mass accelerates only in the -y direction.

   d. The system's center of mass accelerates in both the +x and in the -y direction.

Questions 42-43 refer to the following material.
A block is pulled across a horizontal surface twice. A string attached to the block exerts a force of magnitude F_T on the block at an angle θ_0 above the horizontal in the first run. While the block slides a distance d across the surface, the force of tension exerted on the block by the string does work W_{T1} to the block and the force of friction exerted on the block by the surface does work W_{f1}. The block is returned to its starting position and pulled by the string a second time. In the second run, the string exerts force with the same magnitude F_T, but its direction is changed to be directed at θ_0 below the horizontal. The block again slides a distance d.

42. How does the magnitude of the work done by the force of tension during the second run W_{T2} compare to the work done by the force of tension in the first run?

    a. $W_{T2} > W_{T1}$

    b. $W_{T2} < W_{T1}$

    c. $W_{T2} = W_{T1}$

    d. More information is needed to make the comparison between W_{T2} and W_{T1}

43. How does the magnitude of the work done by the force of friction in the second run W_{f2} compare to the work done by the force of friction in the first run?

    a. $W_{f2} > W_{f1}$

    b. $W_{f2} < W_{f1}$

    c. $W_{f2} = W_{f1}$

    d. More information is needed to make the comparison between W_{f2} and W_{f1}

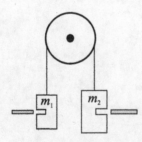

44. Two blocks with masses $m_2 > m_1$ hang by a light string over an ideal pulley as shown above. Each block has a cutout into which a pin can be inserted. When inserted into a block, the pin exerts forces on the block to hold it stationary. Which of the following correctly describes the tension in the string?

    a. The tension in the string will have the same magnitude regardless of the position of the pins.

    b. The tension in the string will be greatest if only the block with mass m_1 has a pin inserted into it.

    c. The tension in the string will be greatest if only the block with mass m_2 has a pin inserted into it.

    d. The tension in the string will be greatest if neither pin is inserted.

45. An object is initially at rest on a level surface. Beginning at time $t_0 = 0$, a constant horizontal force is exerted on the object for 2.00 s. The object slides with negligible friction on the surface. From $t = 0$ to $t = 1.00$ s the work done on the object is 5.00 J. How much work is done on the object from $t = 1.00$ s to $t = 2.00$ s?

   a. 5.00 J

   b. 10.0 J

   c. 15.0 J

   d. 20.0 J

46. In which of the following situations is the net work being done to the object negative?

   a. A crate is moving with constant downward velocity in an elevator.

   b. A ball moves as a projectile along a parabolic path towards its highest point.

   c. A cart moves with negative acceleration while moving with negative velocity.

   d. A dropped leaf accelerates downward with $a < g$ due to a drag force from the air.

47. The equation for the position of an object moving in simple harmonic motion is $x = 2\text{m} \cos\left(\frac{\pi}{2}\frac{\text{rad}}{\text{s}}t\right)$. What is the acceleration a_x of the object at $t = 2\text{s}$?

   a. $\frac{\pi^2}{2}$ m/s^2

   b. $-\frac{\pi^2}{2}$ m/s^2

   c. $\frac{\pi^2}{4}$ m/s^2

   d. $-\frac{\pi^2}{4}$ m/s^2

48. An object is dropped from a height of 20 m above the ground. One second after the first object is released, a second object is thrown downwards, also from a height of 20 m above the ground. How fast must the second object be thrown so that both objects arrive at the ground at the same time?

   a. 5 m/s

   b. 10 m/s

   c. 15 m/s

   d. 20 m/s

49. At time $t = 0$ an object is moving with velocity $-v_0$. A net external force F is then exerted on the object over the interval of time $0 < t < t_1$. At time t_1 the object's velocity is $+v_0$. Which of the following graphs could represent the net force exerted on the object as a function of time?

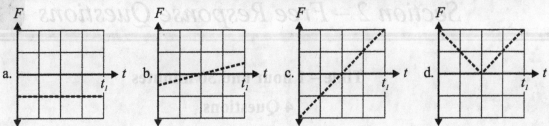

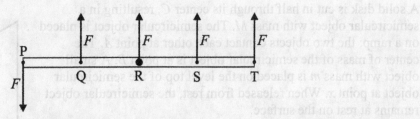

50. A uniform rod is initially at rest but can freely rotate about its center as shown above. Five forces of equal magnitude F are exerted on the rod at equally spaced points P, Q, R, S, and T. At which of the following lettered points could a single force with magnitude $4F$ be additionally exerted perpendicular to the length of the rod so that the rod does not rotate?

   a. P
   b. Q
   c. R
   d. T

Practice Exam I

Section 2 – Free Response Questions

Time – 1 hour and 30 minutes

4 Questions

Directions: Show your work for each of the following in the space provided.

1. A solid disk is cut in half through its center C, resulting in a semicircular object with mass M. The semicircular object is placed on a ramp; the two objects contact each other at point A. The center of mass of the semicircular object is at point B. A small object with mass m is placed on the level top of the semicircular object at point x. When released from rest, the semicircular object remains at rest on the surface.

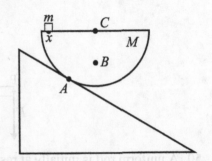

    a. **On the diagram above**, draw and label point P, the center of mass of the system composed of the semicircular object with mass M, and the small object with mass m.

    b. **On the diagram to the right**, draw and label the forces (not components) exerted on the semicircular object as it sits at rest on the inclined surface, which is shown as a dashed line. Draw each force as an arrow starting on, and pointing away from, the point at which the force is exerted. The lengths of the arrows need not indicate the relative magnitudes of the forces.

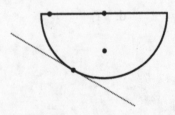

    c. If the small object with mass m is replaced with a small object with mass $2m$, in which direction, if either, will the semicircular object initially roll? Assume that it does not slide on the inclined surface.

        _ up the incline   _ down the incline   _ Neither; it will remain at rest.

        Briefly explain your answer.

2. A lab cart has wheels of negligible mass and can roll with negligible friction. The cart has built-in sensors that can make measurements related to force and motion along the x-axis. The cart's forward direction is defined to be $+x$. The cart can simultaneously transmit any of the following data to a computer:

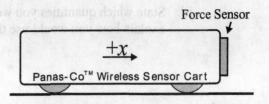

- Position x, measured based on the rolling motion of the wheels with the cart's starting position in each trial being defined to be $x = 0$.
- Velocity v_x, measured based on the rolling motion of the wheels.
- Acceleration a_x, measured by an accelerometer, independent of the wheels.
- Force F_x exerted onto a force sensor at the front end of the cart.

In addition to the cart, you have access to the following equipment. You do not have access to any other equipment.

- A length of track on which the cart can roll in a straight line.
- A computer which can receive data from the cart; data can be displayed directly or as a graph relating any of the quantities measured by the cart's sensors as a function of time, or as a function of another quantity measured by the cart's sensors.

a. When the cart is at rest with all its wheels on the level floor, the accelerometer reports an acceleration $a_x = 0$, but when the cart is at rest with all its wheels against a vertical wall with the front end of the cart facing upward, the accelerometer reports an acceleration $a_x = +9.8$ m/s². Explain why the cart's accelerometer reports this latter acceleration while the cart is at rest held this way.

b. Design an experiment to determine the mass of the cart using only the above equipment to generate a single graph of data measured by one or more of the cart's sensors. Include each of the following in your design:

i. What quantities would be measured and used to determine the mass of the cart?

ii. Give the procedure used to acquire the measurements needed to determine the mass of the cart. Give enough detail so that another student could follow your procedure to conduct one or more additional trials.

Practice Exam I   225

iii. State which quantities you would have the software plot on each of the two axes and explain how you would use this graph to determine the mass of the cart.

The track is now inclined to an angle θ by placing one end of the track on top of a stack of several copies of College Physics for the AP® Physics 1 Course.

c. Design an experiment to determine the angle θ of the inclined track using only the above equipment to generate a single graph of data measured by one or more of the cart's sensors. Include each of the following in your design:

　i. What quantities would be measured and used to determine the angle of the incline?

　ii. Give the procedure used to acquire the measurements needed to determine the angle of the incline. Give enough detail so that another student could follow your procedure to conduct one or more additional trials.

　iii. State which quantities you would have the software plot on each of the two axes and explain how you would use this graph to determine the angle of the incline.

3. The figure to the right shows a block with mass m that slides with negligible friction on a level surface. The block is attached to a spring with spring constant k.

A small ball hangs on the end of a light vertical string with length L directly above the block at position $x = 0$.

The block is pulled to position $x = A$, and the ball is also pulled to position $x = A$, directly above the block such that the string is at an angle θ_0 to the vertical. The block and the ball are both released with zero initial velocity at $t = 0$.

Assume that the angle that the string makes with the vertical is sufficiently small such that the motion of the pendulum is well approximated to be simple harmonic motion.

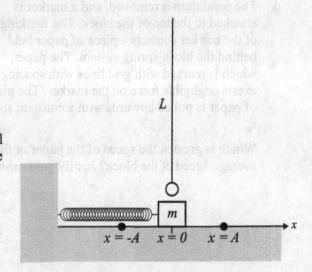

a. Derive an expression for the length of the pendulum L that is needed for the ball to remain directly above the block while they both oscillate. Express your answer in terms of m, k, and the acceleration due to gravity, g.

b. Derive an expression for the speed of the block v_{block} as it moves through position $x = 0$. Express your answer in terms of m, k, and A.

c. A student, in trying to derive an expression for the speed of the ball v_{ball} as it passes through position $x = 0$, writes down the equation $v_{ball} = \sqrt{2gL\cos\theta_0}$. Explain whether or not this equation is plausible (i.e., does it make physical sense?).

d. The pendulum is removed, and a marker is attached to the top of the block. The marking end of the marker contacts a piece of paper held behind the block-spring system. The paper, which is marked with grid lines with spacing D, exerts negligible force on the marker. The piece of paper is pulled upwards with a constant speed v_p.

Which is greater, the speed of the paper or the average speed of the block? Justify your answer.

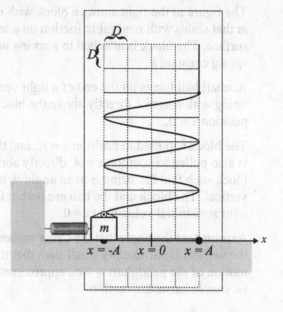

e. The block is placed at position $x = -A$, and a new piece of paper that is identically marked is pulled up with the same constant speed v_p. The block moves from position $x = -A$ to position $x = 0$ at which point it collides inelastically with a block of unknown mass. The two blocks become stuck to each other, and the paper continues to get marked.

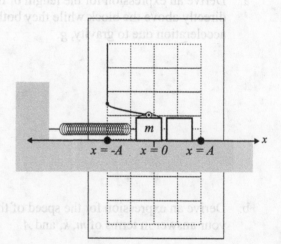

The figure below shows the piece of paper with the mark left on it up to the point of the collision. On the figure below, draw the mark made by the marker as it continues to move until the mark reaches the bottom of the grid.

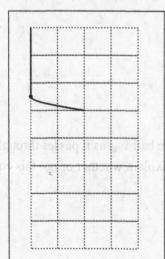

228   Practice Exam I – Free Response Questions

4. A block with volume V_1 and density ρ_1 is fully submerged in a liquid with density $\rho_2 > \rho_1$. A person exerts a force with magnitude F_A to hold the block stationary.

   a. On the dot below, which represents the block while fully submerged and at rest, draw and label the forces (not components) that act on the block. Each force must be represented by a distinct arrow starting on, and pointing away from, the dot.

   b. Derive an expression for the magnitude of the force F_A needed to hold the block fully submerged at rest. Express your answer in terms of V_1, ρ_1, ρ_2, and physical constants, as appropriate.

   c. The block is held stationary with the top and bottom of the block parallel to the surface of the liquid. The top of the block has area A and is a distance d_1 below the surface of the liquid. The bottom of the block also has area A and is a distance d_2 below the surface. Derive an expression for the magnitude of the buoyant force exerted on the block by fluid. Express your answer in terms of ρ_2, A, d_1, d_2, and physical constants, as appropriate.

Practice Exam I   229

d. With the block already fully submerged, the person continues to push the block so that it moves downward in the fluid to a greater depth with a small constant velocity. Assume that the drag force on the block is negligible.

   i. While the block moves downward with constant velocity, does the mechanical energy of the **block-Earth-liquid system** increase, decrease, or remain constant?

   _____ Increase _____ Decrease _____ Remain Constant

   Explain your answer.

   ii. While the block moves downward with constant velocity, does the mechanical energy of the **block-Earth system** increase, decrease, or remain constant?

   _____ Increase _____ Decrease _____ Remain Constant

   Explain your answer.

   iii. While the block moves downward with constant velocity, does the magnitude of the buoyant force increase, decrease, or remain constant?

   _____ Increase _____ Decrease _____ Remain Constant

   Explain your answer.

Practice Exam II
Section 1 – Multiple Choice Questions

Time – 1 hour and 30 minutes
50 Questions

Note: To simplify calculations, you may use $g = 10$ m/s² in all problems.

Directions: For each of the following questions, select the best answer from the four options listed.

1. An elevator is moving upward with constant speed. A person inside the elevator jumps upward. While the person is rising above the floor of the elevator, the person's acceleration is

    a. downward, with a magnitude equal to g.

    b. downward, with a magnitude not equal to g.

    c. upward, with a magnitude equal to g.

    d. upward, with a magnitude not equal to g.

2. An object moves with constant acceleration a for a time t. Which of the following best describes what the product at represents?

    a. The starting velocity of the object.

    b. The velocity of the object at time t.

    c. The average velocity of the object.

    d. The change in the object's velocity.

3. The equation for the position of an object moving in simple harmonic motion is $x = 3\text{m } \cos\left(\dfrac{\pi \text{ rad}}{4 \text{ s}} t\right)$. What is the velocity v_x of the object at $t = 2$s?

    a. $\dfrac{3\pi}{4}$ m/s

    b. $-\dfrac{3\pi}{4}$ m/s

    c. $\dfrac{4\pi}{3}$ m/s

    d. $-\dfrac{4\pi}{3}$ m/s

Practice Exam II   231

4. A satellite moves with a constant speed v_{orbit} in a circular orbit around a planet. Which of the following correctly compares v_{orbit} to the escape speed v_{escape}, the speed needed to escape from the planet, from the satellite's location?

   a. $v_{escape} = v_{orbit}$

   b. $v_{escape} = \sqrt{2}\, v_{orbit}$

   c. $v_{escape} = 2v_{orbit}$

   d. $v_{escape} = 4v_{orbit}$

5. A pendulum is constructed by attaching a small object to a light string which is supported at its upper end. The support does not move. The object is pulled back and released so that it swings back and forth. Is the mechanical energy of the pendulum-Earth system constant?

   a. No, because only the force of gravity does work to the pendulum.

   b. No, because the support at the upper end does work to the system.

   c. No, because the speed of the pendulum is not constant.

   d. Yes, because there are no external forces doing work on the system.

6. Can an object have zero velocity and nonzero acceleration at the same time?

   a. Yes, but only if there is no net force exerted on the object.

   b. Yes, but only for one instant in time.

   c. No, this would violate Newton's second law.

   d. No, this would violate Newton's third law.

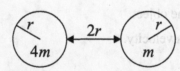

7. The two spheres shown above have the same radii, r, but one sphere has mass $4m$ and the other has mass m. The spheres are separated with a space between their surfaces equal to $2r$. Where is the center of mass of the system composed of the two spheres?

   a. Inside the sphere with mass $4m$.

   b. At the surface of the sphere with mass $4m$.

   c. At a distance $0.5r$ from the surface of the sphere with mass $4m$.

   d. At the midpoint of the line connecting the centers of the two spheres.

8. A 0.80 kg object is projected upward from the floor with an initial speed of 5.0 m/s. 4.0 J of energy is dissipated due to the interaction of the air with the ball while it rises. What is the maximum height reached by the ball?

   a. 0.50 m

   b. 0.75 m

   c. 1.25 m

   d. 1.75 m

9. Four blocks are initially all directly in contact with the floor. Each block is a cube measuring 10.0 cm on each side and has a mass of 1.0 kg. A child then stacks all four blocks vertically on top of each other so that only the bottom block is in contact with the floor. How much work did the child do on the Earth-blocks system?

   a. 0

   b. 4.0 J

   c. 6.0 J

   d. 16 J

Questions 10-12 refer to the following material.

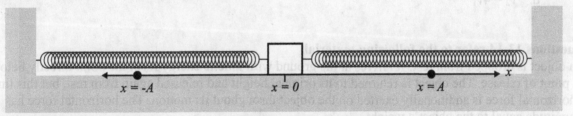

A block with mass m can slide on a level surface with negligible friction. When the block is at position $x = 0$ it is attached to two identical, relaxed springs each with spring constant k as shown in the figure above. The block is then moved to position $x = A$ at which point the spring on the left is stretched and the spring on the right is compressed. The block is released from position $x = A$ at time $t = 0$.

10. How do the forces exerted onto the block by each spring compare when the object is at $x = A$?

   a. They are equal in magnitude and opposite in direction.

   b. They are equal in magnitude and in the same direction.

   c. The magnitude of the force exerted by the spring on the left is greater than the magnitude of the force exerted by the spring on the right.

   d. The magnitude of the force exerted by the spring on the right is greater than the magnitude of the force exerted by the spring on the left.

Practice Exam II   233

11. At what time is the object at position $x = 0$ for the first time?

   a. $t = \dfrac{\pi}{2}\sqrt{\dfrac{m}{2k}}$

   b. $t = \dfrac{\pi}{4}\sqrt{\dfrac{m}{2k}}$

   c. $t = \dfrac{\pi}{2}\sqrt{\dfrac{m}{k}}$

   d. $t = \dfrac{\pi}{4}\sqrt{\dfrac{m}{k}}$

12. What is the speed v_2 of the object when it passes through $x = 0$ when it is being driven by both springs, compared to the speed v_1 it would have passing through $x = 0$ if the spring on the right was removed prior to releasing the object?

   a. $v_2 = v_1$

   b. $v_2 = \sqrt{2}\, v_1$

   c. $v_2 = 2v_1$

   d. $v_2 = 4v_1$

Questions 13-14 refer to the following material.
An object is dropped from rest. It arrives at the ground with kinetic energy K_0 at the point directly below its point of release. The object is returned to its original height and released again from rest, but this time a horizontal force is additionally exerted on the object throughout its motion. The horizontal force has magnitude equal to the object's weight.

13. What is the kinetic energy of the object when it once again arrives at the ground?

   a. K_0

   b. $\sqrt{2}\, K_0$

   c. $2 K_0$

   d. $4 K_0$

14. Which of the following most accurately shows the path the object takes on its way to the ground while the horizontal force is exerted on it throughout its motion?

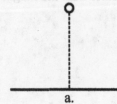
a.
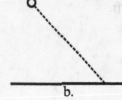
b.
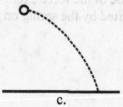
c.
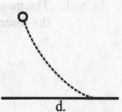
d.

15. Three objects have the same mass. Each object is thrown with the same initial speed from the same height above the ground. Object 1 is thrown vertically upward, object 2 is thrown vertically downward, and object 3 is thrown horizontally. Which of the following correctly ranks the magnitude of the net impulse J provided to each object during its motion to the ground below?

    a. $J_1 = J_2 = J_3$

    b. $J_2 > J_3 > J_1$

    c. $J_1 > J_3 > J_2$

    d. $J_3 > J_1 > J_2$

16. While a door is swinging open, which of the following correctly describes the magnitude of the linear momentum of its center of mass, p, and the magnitude of its angular momentum, L, with respect to the axis passing through the door's hinges?

    a. $p = 0; L = 0$

    b. $p > 0; L > 0$

    c. $p = 0; L > 0$

    d. $p > 0; L = 0$

17. An object slides on a surface with negligible friction towards a spring that is fixed at one end. The object hits the spring which then compresses while slowing the object down until it is momentarily at rest, at which point the spring is compressed 3.0 cm. If the object had been sliding at twice the speed, how much would the spring have been compressed in bringing it momentarily to rest?

    a. 4.2 cm

    b. 6.0 cm

    c. 8.5 cm

    d. 9.0 cm

18. A ball is thrown upwards. Which of the following describes the power delivered to the ball by the force of gravity while the ball is still rising towards its maximum height?

    a. positive with increasing magnitude

    b. positive with decreasing magnitude

    c. negative with increasing magnitude

    d. negative with decreasing magnitude

19. An object is projected horizontally with speed $v < v_{escape}$ from a height of 50 m above the surface of a spherical asteroid that has a radius of 50 km. The only force exerted on the object after being projected is the force of gravity exerted by the asteroid. The possible trajectory of the object's motion can be well-approximated to be what shape(s)?

    a. only parabolic

    b. only circular

    c. only a non-circular ellipse

    d. The trajectory may be well approximated to be any of the above shapes, with the actual shape of the trajectory depending on the magnitude of v.

Questions 20-21 refer to the following material.

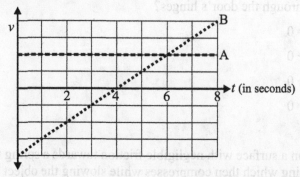

The velocity-time graph above shows the motion of objects A and B which are beside each other at $t = 0$ and move along parallel paths.

20. During which of the following time intervals is the distance between objects A and B decreasing?

    a. Only from $t = 0$ to $t = 4$ s.

    b. Only from $t = 4$ s to $t = 6$ s.

    c. Only from $t = 4$ s to $t = 8$ s

    d. Only from $t = 6$ s to $t = 8$ s.

21. At which of the following times, if any, do the objects have equal displacements from their starting positions?

    a. $t = 2$ s

    b. $t = 6$ s

    c. $t = 8$ s

    d. at no time

22. Two objects of equal mass gravitationally attract each other. If both are released from rest at the same time, they accelerate towards each other, each reaching speed v just before colliding. If instead one of the objects is held stationary and the other is released from rest, what would the released object's speed be just before it collides with the stationary object?

   a. v

   b. $\sqrt{2}\,v$

   c. $2v$

   d. $4v$

23. A cart is initially stationary on a level surface. A person gets the cart moving by exerting a force with magnitude F_0 on the cart for a brief time Δt_0 after which the cart rolls with constant velocity v_x toward a wall, collides with it elastically, and then continues to roll with constant velocity $-v_x$ away from the wall. If the cart was in contact with the wall for the same time Δt_0, what was the magnitude of force that the wall exerted on the cart?

   a. ½ F_0

   b. F_0

   c. $2 F_0$

   d. $4 F_0$

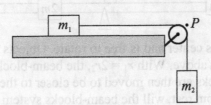

24. Two blocks are attached to a string that goes around an ideal pulley as shown above. Both blocks are released from rest. There is negligible friction between the block with mass m_1 and the surface it is on. Which of the following correctly describes the angular momentum of the two-block system with respect to point P, which is directly to the right of the center of mass of the block with mass m_1 and directly above the center of mass of the block with mass m_2? Consider only the time while the two blocks are moving, before the block with mass m_1 arrives at the pulley.

   a. The angular momentum increases in magnitude and is directed clockwise.

   b. The angular momentum increases in magnitude and is directed counterclockwise.

   c. The angular momentum of the system is constant and zero.

   d. The angular momentum of the system is constant and nonzero.

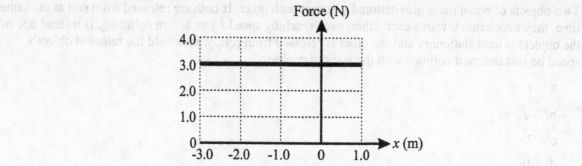

25. A block is sliding in the positive x direction on a level surface with negligible friction. A force is exerted on the block parallel to the x-axis. The graph above shows the net force exerted on the block as a function of position. What is the change in the block's kinetic energy as it moves from $x = -3.0$ m to $x = +1.0$ m?

    a. a decrease of 6.0 J
    b. a decrease of 12 J
    c. an increase of 6.0 J
    d. an increase of 12 J

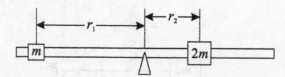

26. A light beam is supported at its center and is free to rotate. Objects with masses m and $2m$ are attached to the beam as shown above. With $r_1 = 2r_2$, the beam-blocks system remains stationary when released from rest. If both blocks are then moved to be closer to the center by moving them the same distance $d < r_2$ and released from rest, will the beam-blocks system remain stationary?

    a. Yes, the system will remain stationary.
    b. No, the system will move; the block on the left will initially rise and the block on the right will initially move downward.
    c. No, the system will move; the block on the left will initially move downward and the block on the right will initially rise.
    d. No, the system will move; even with $d < r_2$ the direction of the initial motions of the blocks depends on the size of d.

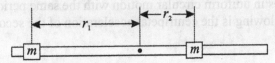

27. A light beam is supported at its center and is free to rotate. Two objects each with mass *m* are attached to the beam as shown above. When the beam-blocks system is released from rest, will the beam-blocks system remain stationary?

    a. Yes, the system will remain stationary.

    b. No, the system will rotate with constant angular speed.

    c. No, the system will rotate with a constant non-zero angular acceleration.

    d. No, the system will rotate with an angular acceleration that changes with time.

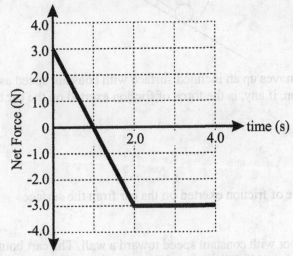

28. An object initially has zero velocity on a surface with negligible friction. The graph above shows the net force exerted on the object as a function of time. When is the net force doing negative work?

    a. from 0 to 1.0 s only

    b. from 1.0 s to 2.0 s only

    c. from 0 to 2.0 s only

    d. from 1.0 s to 4.0 s only

29. A ball is rolling on the ground to the east. The ball hits a tree and as a result begins to roll to the north. Which of the following best describes the direction of the impulse that the ball exerted on the tree?

    a. northeast

    b. southeast

    c. northwest

    d. southwest

30. An object moves in uniform circular motion with period T, speed v_1, and has centripetal acceleration a_1. A second object moves in uniform circular motion with the same period T but moves at speed $v_2 = 2v_1$. Which of the following is the centripetal acceleration of the second object a_2 in terms of a_1?

    a. $a_2 = \frac{1}{2} a_1$

    b. $a_2 = a_1$

    c. $a_2 = 2 a_1$

    d. $a_2 = 4 a_1$

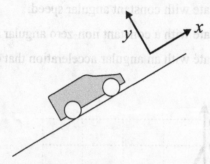

31. A battery-powered car moves up an inclined surface with constant speed as shown in the figure above. In which direction, if any, is the force of friction exerted on the car by the surface?

    a. $+x$

    b. $-x$

    c. $+y$

    d. There is no force of friction exerted on the car from the surface.

32. A cart rolls along the floor with constant speed toward a wall. The cart bounces off the wall, reversing its direction of motion in a negligible amount of time. It then rolls back to its starting position with a smaller, but again constant speed. Which of the following position-time graphs best shows the motion of the cart?

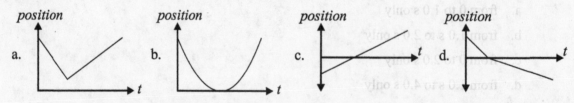

33. Four small objects each with mass m are positioned one each at the four corners of a square with side length L. What is the rotational inertia of the system composed of the four small objects with respect to the axis that passes through one of the corners of the square?

    a. mL^2

    b. $2 mL^2$

    c. $3 mL^2$

    d. $4 mL^2$

34. A 2.0 kg object is initially at rest when it is struck by an 8.0 kg object. The two objects move together at 5.0 m/s immediately after the collision. For the system composed of the two objects, what was the velocity of the system's center of mass immediately before the collision?

    a. 5.0 m/s
    b. 6.3 m/s
    c. 25.0 m/s
    d. 50.0 m/s

35. A light suction cup clings to a smooth ceiling, supporting the weight of a block that is attached to the suction cup. Which one of the following statements is necessarily true?

    a. The upward force from the air in the room exerted on the suction cup is greater than the weight of the block.
    b. The upward force from the air in the room exerted on the suction cup must be equal to the weight of the block.
    c. The air inside the suction cup pulls the suction cup upwards with a force equal to the weight of the block.
    d. The air inside the suction cup pushes the suction cup downwards with a force equal to the weight of the block.

36. The continuity equation is based on which law of Physics?

    a. Conservation of linear momentum
    b. Conservation of mass
    c. Conservation of energy
    d. Newton's 1st law

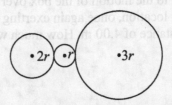

37. Three wheels with radii r, $2r$, and $3r$ are each free to rotate about its center. The smallest wheel is in contact with both other wheels and rolls against them without sliding. If the wheel with radius $2r$ is rotating with angular speed ω, what is the angular speed of the wheel with radius $3r$?

    a. $\omega/3$
    b. $2\omega/3$
    c. ω
    d. $3\omega/2$

38. A person stands on the floor of a bus while the bus moves around a level circular curve with constant speed. The person maintains their balance on the floor without sliding. Which of the following forces is directed away from the center of the circular path?

    a. Static friction force exerted on the person by the floor.

    b. Kinetic friction force exerted on the person by the floor.

    c. Static friction force exerted on the floor by the person.

    d. There is no force directed away from the center of the circle.

39. A horizontal 4 cm diameter pipe tapers to a 2 cm nozzle. If the water emerges from the nozzle at atmospheric pressure (101.3 kPa) with a speed of 20 m/s, what is the pressure in the 4 cm section of the pipe?

    a. 107.5 kPa

    b. 190 kPa

    c. 250 kPa

    d. 290 kPa

40. A fluid with density ρ flows with speed v out of a hole with area A. Which of the following is an expression for the rate at which kinetic energy is transferred through the hole?

    a. $\rho A v^2$

    b. $\rho A v^3$

    c. $\frac{1}{2} \rho A v^2$

    d. $\frac{1}{2} \rho A v^3$

41. A person pushes a box across a rough floor from one end of a room to the other end, exerting a horizontal force of 15.0 N parallel to the motion of the box over a distance of 4.00 m. The person then pushes the box back to its original location, once again exerting a horizontal force of 15.0 N parallel to the motion of the box over a distance of 4.00 m. How much work did the person do to the box in total?

    a. 0

    b. 60.0 J

    c. 120.0 J

    d. The work cannot be determined without knowing the mass of the box.

42. A 2.0 kg object is initially at rest. A net force that gradually increases is then exerted onto the object. The net force is initially zero but increases at a uniform rate of 4.0 N/s. What is the speed of the object after 6.0 s?

   a. 12 m/s
   b. 24 m/s
   c. 36 m/s
   d. 72 m/s

43. A small object with mass m moves back and forth between two fixed blocks which are a distance L apart. The only force exerted on the object is due to collisions with the blocks. The object moves with constant speed v between the blocks and reverses its direction in a negligible amount of time each time it hits a block. Averaged over a large amount of time involving many collisions, which of the following is an expression for the magnitude of the average force the object exerts on one of the blocks?

   a. $\dfrac{mv}{L}$
   b. $\dfrac{mv}{2L}$
   c. $\dfrac{mv^2}{L}$
   d. $\dfrac{mv^2}{2L}$

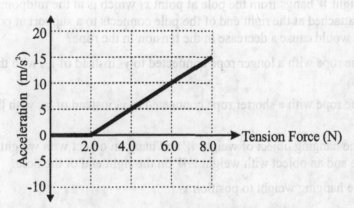

44. A box is initially at rest on the surface of a faraway planet. A rope is tied to the box so that it can lift the box up. The tension in the rope begins at zero and is then slowly increased. The graph above shows the acceleration of the box as a function of the tension in the rope. Which of the following is closest to the acceleration due to gravity on the planet the box is on?

   a. 2.0 m/s²
   b. 2.5 m/s²
   c. 5.0 m/s²
   d. 15 m/s²

45. Circular objects X and Y have the same mass. The two objects are released with zero initial velocity from an inclined surface. Both objects roll without sliding to the bottom of the incline, with object X reaching the bottom before object Y. Object X is returned to the top of the incline, but object Y is placed only partway up the incline. Both objects are again released with zero initial velocity. Both objects roll without sliding and reach the bottom of the incline at the same time. How do the objects' rotational inertias I and kinetic energies K when they reach the bottom of the incline at the same time compare to each other?

   a. $I_X < I_Y; K_X > K_Y$

   b. $I_X < I_Y; K_X = K_Y$

   c. $I_X > I_Y; K_X > K_Y$

   d. $I_X > I_Y; K_X = K_Y$

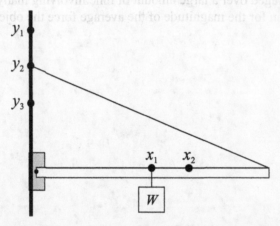

46. The figure above shows a horizontal uniform pole which is free to rotate about a hinge at its left end. An object with weight W hangs from the pole at point x_1 which is at the midpoint of the length of the pole. A light rope attached at the right end of the pole connects to a support at point y_2. Which of the following changes would cause a decrease in the tension in the rope?

   a. Replace the rope with a longer rope connected to y_1 instead of y_2, with the pipe remaining horizontal.

   b. Replace the rope with a shorter rope connected to y_3 instead of y_2, with the pipe remaining horizontal.

   c. Remove the hanging object of weight W and hang an object with weight $½W$ on the left end of the pole and an object with weight $½W$ on the right end of the pole.

   d. Change the hanging weight to position x_2.

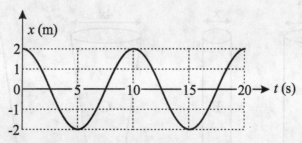

47. The position-time graph for an object moving in simple harmonic motion is shown above. Which of the following is an equation for the position of the object?

   a. $x = (2 \text{ m})\cos\left(\dfrac{\pi \text{ rad}}{10 \text{ s}} t\right)$

   b. $x = (2 \text{ m})\cos\left(\dfrac{\pi \text{ rad}}{5 \text{ s}} t\right)$

   c. $x = (4 \text{ m})\cos\left(\dfrac{\pi \text{ rad}}{10 \text{ s}} t\right)$

   d. $x = (4 \text{ m})\cos\left(\dfrac{\pi \text{ rad}}{5 \text{ s}} t\right)$

48. A block is placed on a surface for which $\mu_S = 0.6$ and $\mu_k = 0.4$. At which one of the following angles could the surface be inclined so that the block can remain at rest on the surface, but if it is given a brief push by your hand, the block could be sent sliding down the ramp with increasing speed, even when it is no longer being pushed by your hand?

   a. 15°
   b. 25°
   c. 35°
   d. 45°

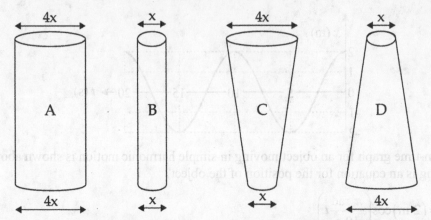

49. Four light containers with equal heights but different shapes are filled with water as shown above. Which of the following correctly ranks the pressures associated with the force that the container exerts onto the surface it is at rest on?

   a. $P_A = P_B = P_C = P_D$

   b. $P_C < P_A = P_B = < P_D$

   c. $P_B < P_C = P_D < P_A$

   d. $P_D < P_A = P_B < P_C$

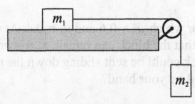

50. Two objects m_1 and $m_2 > m_1$ are attached to a string which passes over an ideal pulley as shown here. The object with mass m_1 slides to the right with increasing speed across the surface which exerts negligible friction force on it. Which of the following changes would result in a reduction in the tension in the string?

   a. Replacing the object on the surface with an object that has less mass than the original object.

   b. Changing the orientation of the surface so that the end with the pulley is higher than the other end.

   c. Changing the surface to one that exerts a friction force on the object as it slides.

   d. Exerting a force to the left on the object with mass m_1 that prevents it from moving.

Practice Exam II
Section 2 – Free Response Questions

Time – 1 hour and 30 minutes

4 Questions

Directions: Show your work for each of the following in the space provided.

1. The figure here shows an elliptical orbit of a satellite around a planet that has much greater mass than the satellite. The satellite orbits in the counterclockwise direction.

    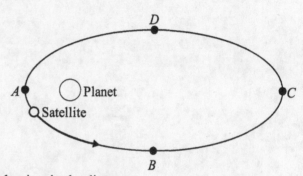

    a. Does the force of gravity exerted onto the satellite from the planet ever do positive work to the system composed of only the satellite? If so, indicate all segments of the orbit for which this work is done, with reference to the lettered points in the diagram.

    Justify your answer.

    b. Does the force of gravity exerted onto the satellite from the planet ever do negative work to the system composed of only the satellite? If so, indicate all segments of the orbit for which this work is done, with reference to the lettered points in the diagram.

    Justify your answer.

    c. As the satellite moves from point B to point C, is the mechanical energy of the system composed of the satellite and the planet increasing, decreasing, or constant?

    _____ Increasing   _____ Decreasing   _____ Constant

    Justify your answer.

2. A ball is dropped at $t = 0$. The ball falls in the presence of air resistance, reaching terminal speed at $t = t_1$, and then continues to fall at terminal speed until $t = t_2$.

   a. Sketch a position-time graph that shows the motion of the ball from $t = 0$ until $t = t_2$. Label t_1 and t_2 on the time axis.

   b. Sketch a velocity-time graph that shows the motion of the ball from $t = 0$ until $t = t_2$. Label t_1 and t_2 on the time axis.

c. Two students are discussing how a spherical raindrop falls while a force is exerted onto it from the air. The students conduct an internet search and find an equation for the terminal speed of a spherical falling raindrop in the presence of air: $v_{terminal} = \sqrt{\dfrac{2mg}{kA}}$ in which m is the mass of the raindrop, g is the gravitational field strength, A is the cross-sectional area of the sphere, and k is a positive constant.

One of the students concludes that the equation they found for terminal speed doesn't make sense because it suggests that terminal speed decreases as area A increases. The student correctly points out that a larger raindrop falls with a greater terminal speed than a smaller raindrop. Is the student's conclusion correct? Justify your answer.

3. The figure below shows two objects with masses m_1 and $m_2 > m_1$ attached by a string with negligible mass which passes over a pulley which is free to rotate. Both objects are released with zero initial velocity. Assume that the string does not slip on the pulley.

If the pulley has negligible mass, the string at point A is under tension F_{A1} and the string at point B is under tension F_{B1}. If instead, the pulley has mass that is not negligible, the string at point A is under tension F_{A2} and the string at point B is under tension F_{B2}.

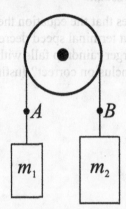

a. The diagrams below represent the pulley in each of the two cases described above. On each diagram, draw and label the forces (not components) that are exerted on the pulley. Draw each force as an arrow starting on, and pointing away from, the point at which the force is exerted. The lengths of the arrows need not indicate the relative magnitudes of the forces.

b. How does the magnitude of F_{A1} compare to the magnitude of F_{B1}?

_____ $F_{A1} < F_{B1}$ _____ $F_{A1} > F_{B1}$ _____ $F_{A1} = F_{B1}$

Justify your answer.

c. How does the magnitude of F_{A2} compare to the magnitude of F_{B2}?

_____ $F_{A2} < F_{B2}$ _____ $F_{A2} > F_{B2}$ _____ $F_{A2} = F_{B2}$

Justify your answer.

4. A small 2 kg ball is supported at rest by two light strings of equal length. Each string has one end fastened to the ball and the other ends of the strings are attached to a horizontal rod as seen below.

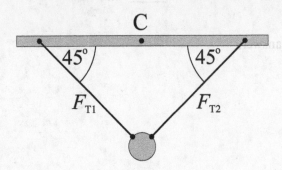

    a. Calculate the magnitude of the tension in the left string F_{T1}.

The rod is then made to rotate at a uniform rate around its center C, resulting in the ball moving in a vertical circle centered on C with radius 0.8 m. At one moment the rod is vertical as seen below.

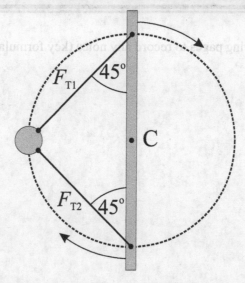

b. At the moment shown above, how does the magnitude of the tension in the upper string F_{T1} compare to the magnitude of tension in the lower string F_{T2}? Explain your answer.

c. Calculate the angular speed the rod must have so that the ball maintains motion at a constant speed, with the tension in both strings momentarily becoming zero when the ball is at the highest point in its motion.

Notes

Please feel free to use the following pages to record any notes (key formulas, equations, diagrams, etc.) which will be helpful for review.

Notes

Notes

Notes

Notes

Notes

Notes